THE HUMAN, THE ORCHID, AND THE OCTOPUS

THE HUMAN, THE ORCHID, AND THE OCTOPUS

EXPLORING AND CONSERVING OUR NATURAL WORLD

Jacques Cousteau

AND SUSAN SCHIEFELBEIN

BLOOMSBURY

Published by Bloomsbury USA, New York
Distributed to the trade by Holtzbrinck Publishers

All papers used by Bloomsbury USA are natural, recyclable products
made from wood grown in well-managed forests. The manufacturing
processes conform to the environmental regulations of the country of origin.

LIBRARY OF CONGRESS CATALOGING-IN-PUBLICATION DATA

Cousteau, Jacques Yves.
The human, the orchid, and the octopus : exploring and conserving our
natural world / Jacques Cousteau and Susan Schiefelbein.
p. cm.
ISBN-13: 978-1-59691-417-9 (hardcover)
ISBN-10: 1-59691-417-3 (hardcover)
1. Cousteau, Jacques Yves.   2. Oceanographers—France—Biography.
I. Schiefelbein, Susan.   II. Title.
GC30.C68A3 2007
333.95'16—dc22
2007018824

First U.S. Edition 2007

1 3 5 7 9 10 8 6 4 2

Typeset by Westchester Book Group
Printed in the United States of America by Quebecor World Fairfield

CONTENTS

. . . I am a part of all that I have met;
Yet all experience is an arch wherethro'
Gleams that untravell'd world whose margin fades
For ever and forever when I move . . .

. . . .

The lights begin to twinkle from the rocks;
The long day wanes: the slow moon climbs: the deep
Moans round with many voices. Come, my friends,
'Tis not too late to seek a newer world . . .

"Ulysses"
ALFRED, LORD TENNYSON

FOREWORD

By Bill McKibben

FOR those of us who came of age in the 1960s or '70s, the picture of Jacques Cousteau is fixed forever in our minds. A slight but wiry man, yellow tank peeking over his shoulder, falling backward off the stern of the good ship *Calypso* as he prepared for yet another dive down among the rays or the jellyfish or the sea cows or the parrot fish— down, literally, into his world, "the undersea world of Jacques Cousteau." His voice became just as familiar, with its somehow slightly wistful but still infectious Gallic intonation. "In ze wisdom of ze dolphins lies ze test of human wisdom." He was as commanding a media persona as Walt Disney (who had his own fantastical world); there were very few foreigners that Americans viewed with more complete trust and admiration.

And why not? As this rich new book reminds us, he was utterly trustworthy, a figure, like Rachel Carson, moved by no desire deeper than to appreciate the world around him, to share that love, and thus to protect it. He was the quintessential explorer—but whereas most great explorers come equipped with a kind of maniacal drive that allows them to scale huge mountains or beat their way across untracked jungle (and also to fight endlessly about who got there first), there was always something more peaceful about Cousteau. Maybe it was a lifetime of swimming off coral reefs—that most enchanted corner of God's brain, where the animals seem hardly to notice that you're there, and barely bother to get out of your way. The places and creatures he showed us might be dangerous at times, but mostly his world was benign, often sun-dappled, incredibly attractive.

Not only that, but it was brand new. These were the days before many people went scuba diving (partly because Cousteau had just gotten around to inventing much of the necessary equipment). Since we're terrestrial creatures, and since you usually can't see very far down into the oceans, they were for most of us a kind of blank—our sense of them came largely from what you pulled up with a net from their depths, or what washed up on the beach. Neil Armstrong may have taken us to the moon, but the journey Cousteau took us on was very nearly as shocking.

And much more important. Because as we were slowly starting to realize, the timeless sea, like everything else on Earth, was under assault. Cousteau arrived on the American scene at the first great ecological moment—when *Silent Spring* and David Brower and the burning Cuyahoga River and the first Earth Day were all combining to make us aware that the planet was fragile, and that what's more we were breaking it. The oceans had always seemed unfathomably (pun intended) vast. Indeed, generations of engineers had learned the catechism "The solution to pollution is dilution" and taken it as a license to dump stuff beneath the waves. But all of a sudden that stuff was starting to change life underseas in profound ways. Cousteau and his team found evidence of toxic contamination in many of the places where they dived, and evidence too of overdevelopment and overfishing.

Cousteau divided his career, therefore, between two tasks, equally necessary: getting people to marvel at the beauty of the oceans, and then pointing out how we were destroying them. It was as if the earliest explorer of the North American continent was simultaneously cataloging its vast buffalo herds and watching them die, or if the earliest trips up the Amazon were uncovering the diversity of wildlife while also watching large-scale clear-cutting. No explorer has ever been faced with quite such a dilemma, and Cousteau handled it superbly.

On the one hand he was tough—the same steeliness that made him effective on behalf of the Resistance during the war made him a man to reckon with on matters environmental. (The tale of his confrontation with de Gaulle over the French nuclear program is by itself worth the price of the book.) He was willing to confront scientists who didn't stand up for the seas they studied, and fishing tycoons, and navies. But on the other hand he was seductive—he gave people the chance to feel at home in a beautiful and mysterious world, and the chance to help safeguard it. Someone once said of the American nature writer John Burroughs that he had, with his hiking tales, launched a thousand backpacks. Cousteau launched a million pairs of flippers, and many thousands of activist careers.

Whether those activists will succeed in saving the world he opened to our view is very much an unanswered question. When he died in 1997, the greatest threats to our marine ecosystems were only beginning to heave into view. Global warming, for instance, now threatens to destroy all the earth's coral reefs before the century is out—the small animal that creates the reef can't survive outside a fairly narrow band of temperatures. Meanwhile, great heat-spawned storms like Katrina endanger coastlines, and the inexorable rise in sea level seems destined to wreck many of the ocean-fringing marshes and wetlands that are the sea's biological batteries. In the past year or two, scientists have begun to warn of an even more profound danger. The same carbon dioxide warming the atmosphere has begun to dramatically alter the pH of the seas, in ways that marine life simply may not be able to cope with. We face not just a silent spring but an empty ocean.

If we are to meet those challenges, a healthy dose of fear will serve as one motivator. Cousteau was insistent throughout his life that this was a water planet—that the thing that made Earth distinctive was its vast liquid seas, against which our more familiar terra firma was in some sense mere backdrop. We came from the seas, and without them our survival was impossible.

But fear is the second-best motivator. Love tops it, and love is what Cousteau specialized in. Of all the people who lived inside our television sets in the early decades of the screen age, none was more purely in love with the planet than Jacques Cousteau. He took us places we'd never been before and shared their overwhelming beauty. He set us on a great adventure. We can only hope we'll prove equal to his challenge.

INTRODUCTION

By Susan Schiefelbein

I'D been traveling upriver on the Amazon for about a week when Jacques-Yves Cousteau suggested that I leave his ship *Calypso* to join a land team driving into the interior. He'd asked me to write the narration for a documentary he was producing on the Amazon basin, and he wanted me to accompany his men across Brazil to its border with Bolivia, where impoverished villagers were attempting to dive for gold buried in the bottom sands of one of the river's tributaries. His men and I clambered into a truck and set off onto a "highway" that was in fact nothing more than a dirt road into the rain forest. The trail became ever more narrow as the undergrowth around us became more dense. After proceeding for hours deep into the jungle, the driver suddenly stopped. We jumped down from the truck and instantly understood our dilemma. A massive tree lay prone across the trail ahead. Even lying on its side, it came up to our waists. None of us spoke. We'd all heard a story about Yanomamo Indians who had reputedly killed a busload of tourists for having trespassed on tribal territory via a so-called highway like ours. The tree before us had obviously been felled intentionally. It was a warning. No question of trying to move it and continue on. The men had to find a way somehow to turn the truck around on the narrow trail and get out. As we stood in silence, the thick foliage on one side of the path parted. A brown-skinned man stepped out. His eyes settled on me, the only woman. He pointed to my heart. In the same split second that I realized I was wearing a Cousteau Society T-shirt, he burst into a broad grin and sang out—in English—"Jacques Cousteau! Way to go!"

I believe I can speak for the group in saying that none of us was really surprised to learn that Cousteau's fame had penetrated even into the depths of the most expansive rain forest on Earth. Wherever and however this man had seen a television, the fact remained that at the time, each of Cousteau's documentaries was attracting some 250 million viewers distributed across almost every country in the world. Cousteau was said to have turned the television screen into a porthole, and people around the globe eagerly gazed through it to catch a glimpse of the grandfatherly figure who, with his red watch cap and wire-rimmed spectacles, spoke about the wonders of the undersea world in his lyrical French accent.

Yet while millions of televiewers knew Cousteau's documentaries, I always wished that more knew his true vocations: exploration, invention, and civic duty. He was gifted with a creative intellect that so inspired NASA and the National Geographic Society's historic exploration committee—as well as France's Navy, its Ministry of Education, and its Natural History Museum—that they invested millions in his inventions and expeditions. They paired him with the noted engineers, scientists, and photographers of the day. The partnerships produced innovations that changed the face of the twentieth century. With heads of state and heads of international corporations eagerly seeking his counsel, Cousteau was uniquely positioned to comment not just on the marine world but also on the world at large. He was much more than a filmmaker. He was one of the first true global citizens.

Cousteau never ceased to appreciate the fact that he'd been born at what he called his "lucky place in time"—exactly at the moment, as he described it, when new technology could expedite the work of a lone explorer but before advanced technology made the lone explorer obsolete. In fact, Cousteau made his own luck. He took full advantage of every experience that came his way. His father, adviser to international millionaires, traveled constantly with his wife and children in tow. By the age of ten, Jacques-Yves was learning English while spending a year with his family in the United States. As a thirteen-year-old back in France, he proudly accepted his father's gift of a 9 mm Pathé Baby camera and began making amateur movies in which he served as writer, director, and star. He even taught himself to develop film with chemicals he pinched from his grandfather's pharmacy. Next on his list: swimming. A sickly child suffering from enteritis and a heart murmur, the young Jacques-Yves had been encouraged to strengthen himself through exercise. Before long he became a national university champion.

After graduating from the Collège Stanislas—attended a few years before by Charles de Gaulle—Cousteau applied to France's Naval Academy, one of the most demanding academic institutions in the country. In 1933 he graduated second in his class. "I even studied with a flashlight in bed," he later remembered. "I wanted the two greatest things the Navy had to offer me: discipline and a formidable education in technology."

The Navy, of course, also gave Cousteau the opportunity to develop the global perspective so evident in the anecdotes recounted in this book. As a young midshipman, he took the traditional world tour on the training ship *Jeanne d'Arc*. Slung over his shoulder was his ever-present moving-picture camera, with which he recorded his ports of call, including Bali and even Hollywood. He next boarded the *Primauguet*, a cruiser that sailed to the Far East, where he commanded—and of course filmed—the French naval base at Shanghai.

On his return to France, Cousteau enrolled in the Navy's School of Maritime Aviation. His steps toward an aeronautical career came to an abrupt halt when a catastrophic car crash resulted in serious injuries that compromised the use of both of his arms. Cousteau turned what would have been the ultimate bad fortune for anyone else—the end of his dreams of becoming a pilot—into the luckiest event of his life. In 1936, at the French naval base in the Mediterranean city of Toulon, he once again began swimming to strengthen himself. A familiar figure on the beaches of southern France and at Navy seminars regarding underwater weapons, he struck up a friendship with a superior officer, Commandant Philippe Tailliez, and later with Tailliez's friend Frédéric Dumas. Known as "Didi," Dumas had gained a reputation as a quasi-miraculous champion in spearfishing, a sport that in the 1930s had become all the rage along the Mediterranean.

Today it is difficult to imagine how, such a relatively short time ago, human beings could have perceived the deep sea so differently—as nothing much more than a vast, glittering surface. What existed in the dark bottom waters, no one knew. In the popular imagination, the demonic depths swallowed drowning men and titanic ships alike, pulling them down into an abyss swept by monsters the likes of which had sprung from the minds of men like Jules Verne and Victor Hugo.

At the time, the common method of diving remained as primitive as it had been for millennia: by apnea, that is, holding one's breath and performing a surface dive. The first pair of flippers had been patented only three years before Cousteau had been posted to Toulon. The blur and sting of saltwater on divers' eyes compromised vision. All this would soon change for Cousteau thanks to one simple gesture from a friend. Tailliez lent him a pair of goggles used by Philippine pearl fishermen. "I'd navigated the seven seas and I'd swum in all of them, but it was as a blind man that I made those first contacts," Cousteau remembered. "In 1936, only, that first pair of underwater goggles permitted me to open my eyes on an unknown domain. From that moment on I never looked back. With my friends Tailliez and Dumas, I dived year-round, in warm and icy waters. I learned how to dive down sixty feet and harpoon a fish in a few seconds."

The Cousteau-Tailliez-Dumas trio became such a fixture along the rocky inlets of Toulon that before long the friends became known not as the three *mousquetaires* but as the three *mousque-mers*—a rhyming play on the French words for the three "musketeers of the sea." They were soon joined by a fourth. Cousteau married the seventeen-year-old Simone Melchior. A glance at the petite Simone could have easily misled anyone who doubted her stalwart seagoing heritage: On her mother's side, she was both a grandchild and a great-grandchild of admirals. Her paternal grandfather was none other than Vice Admiral Jules-Bernard-François Melchior, so respected a name in French naval history that the legendary explorer Jean-Baptiste Charcot, who sailed his renowned ship the *Pourquoi Pas?* to the Antarctic, named the Melchior Islands after him. In an era when military careers for women were nonexistent, Simone wanted nothing more than to become a Navy officer. She liked to say she did the next best thing—she married one.

World War II interrupted the quartet's activities, but it also provided Cousteau with the experiences that one day not only would earn him the respect of government leaders but also would shape his advocacy for peace and his convictions about the need for an effective United Nations. Cousteau served on the French cruiser *Dupleix*, hunting down the *Graf von Spree*, one of the small but heavily armed battleships that Germany had introduced into the war. For his participation as gunnery officer in the bombardment of Genoa he was awarded the Croix de Guerre with distinction and two citations. When Germany occupied the north of France, the French Navy posted Cousteau back on the southern coast as an intelligence agent. As described in the second chapter of this book, he staged a burglary into Italian headquarters worthy of Hitchcock, and moreover he managed to loosen the tongues of officers of the German Navy—after generously offering them rounds of drinks—thus obtaining the information that Hitler wanted to seize the French fleet. Although Cousteau managed to convince his superiors of the reality of this threat, he was crushed when the powers-that-be in the free south of France responded by destroying their own ships. Cousteau was later awarded the Légion d'honneur, but for exactly what actions neither he nor anyone else in the French intelligence community ever detailed. The official journal published at the time cryptically announced: "Cousteau, Jacques-Yves, ship's lieutenant: during the war, rendered eminent services in circumstances particularly perilous."

Yet even during these stresses of wartime, Cousteau found a way to be creative and productive. What better cover for his intelligence work—for explaining his constant vigilance over the coastline—than by giving himself up to his passion for diving and filming with Tailliez and Dumas? Ever the inventive handyman, Cousteau made his camera watertight by placing it in a glass canning jar. "When we filmed, the Germans often boarded our support boat," remembered Tailliez. "We would give them goggles and tell them to look underwater. They'd say, '*Schoen, schoen*,' beautiful, beautiful. It was bizarre."

Diving only by apnea with his friends, Cousteau made his first short film about his dives to sixty feet, which he poetically titled *Ten Fathoms Down*. He longed for more. "Always I rebelled against the limitations imposed by a single lung-full of air," he recalled. "In lightning dives, we could only get a glimpse of the sea's enigmas."

Cousteau's pioneering efforts to dive deeper and stay longer would bring him face to face with greater hazards than even novelists like Jules Verne had imagined. He and his team confronted these dangers in a series of trial-and-error experiments, many of which became the basis for the rules of modern undersea diving.

Cousteau attempted to dive with the Fernez apparatus, at the time a new invention in which pressurized air was pumped from tanks on a support boat via a tube to a full face mask worn by the diver below. As Cousteau described it, he was languidly swimming in the depths when he suddenly felt a shock in his lungs. The tube leading from the support ship had broken somewhere up above the surface. He barely escaped drowning: "If I hadn't instinctively shut the glottis in my throat, the broken tube would have fed me thin surface air and the weight of the water would have collapsed my lungs."

Cousteau again nearly lost his life while experimenting with a "rebreather," another device new at the time that fed the diver pure oxygen. Accepting the information of the day—that pure oxygen does not become toxic until forty-five feet down and not, as Cousteau unfortunately was to discover, at about thirty-three feet—he was reveling in a dive when, as he remembered it, "my lips began to tremble uncontrollably. My eyelids flutttered. My spine bent backward like a bow." Before he lost consciousness, he somehow managed to unfasten his weight belt. The sailors he'd brought along to spot him from above saw his inert body float to the surface and fished him out of the water. Modern medical journals have referenced Cousteau's experiments; the problems associated with breathing pure oxygen at certain depths are associated with those of breathing pure oxygen for prolonged periods for medical purposes.

About this time, the French Navy officer Commandant Yves Le Prieur came extremely close to creating the device of Cousteau's dreams. Le Prieur's invention consisted of a single tank, worn on the back, that released a constant flow of air. The diver regulated pressure manually by means of a valve on the tank. Unfortunately, the uninterrupted airflow emptied the tank quickly and limited divers to only brief stays underwater. Cousteau himself credited Le Prieur with paternity for the modern underwater breathing device. The reason why it was Cousteau's name, and not Le Prieur's, that entered the lexicon lay in the simple fact that Cousteau wanted more. He dreamed of flying through water—hovering in the three-dimensional world like a fish—without the preoccupation of fidgeting with a valve that moreover created a drag in water. Cousteau envisioned an apparatus—a demand valve, a "regulator"—that would supply air only on a diver's inhalation, shut off on exhalation, and regulate air pressure automatically as a diver descended.

The youthful Simone Cousteau stepped in. While many of her ancestors were men of the sea, her father was a leading executive of Air Liquide, at the time the largest supplier of natural gas in France. In December 1942, Simone's father introduced Cousteau to Émile Gagnan, one of the company's most brilliant engineers. Cousteau described his idea for a regulator that would feed a diver air on demand at the ambient pressure of the sea. Gagnan cheerfully tossed a device onto a table and asked, "Something like this?" Gagnan had already invented a regulator with which cars could run on cooking gas during the wartime years of petroleum shortages.

Cousteau spent months with Gagnan, outlining exactly what he wanted, and together the two adapted the regulator to diving. In spring 1943, Cousteau donned a pressurized air tank outfitted with the adapted regulator and plunged into the Marne River outside Paris. When he was upright, air overflowed; when he headed downward, the air flow ceased. Back to the drawing board. The war itself created delays. Cousteau had built three prototypes of the breathing apparatus—one for himself as well as one each for his fellow experimenters Dumas and Tailliez. He reportedly had to produce a fourth when his own, set aside on a beach, was blown to bits by a misfired Allied artillery shell. Still the three friends persisted and together made five hundred separate test dives. By autumn 1943, Dumas had achieved an attested descent down to 220 feet; Cousteau himself later atttained a depth of 333 feet—tremendously hazardous ventures into the unknown, especially when one considers that today many agencies certify even advanced sports divers on normal air down only to

80 to 150 feet. By 1944 Simone would become the first female diver with the device they'd all christened the "Aqualung." Soon Cousteau's two sons, aged seven and four, would dive with child-sized gear. When the children choked on water because they couldn't refrain from exclaiming at the wonders they saw, their father admonished them by saying that the underwater realm is a "silent world." In 1946 Cousteau and Gagnan patented their invention.

Today, the principle of the Cousteau-Gagnan regulator remains unchanged. The four prototypes that Cousteau originally produced have proliferated into untold millions of the device popularly known as the SCUBA (self-contained underwater breathing apparatus)—with an estimated forty thousand neophyte divers now being certified worldwide each month. It has often been repeated—including by Cousteau—that one of the greatest moments in modern times occurred when the moon astronauts looked back from the void of space and saw the "beautiful wet blue ball" of our water planet. Yet even the astronauts lay eyes only on an azure surface. It was Jacques-Yves Cousteau who opened its portals and invited our species to enter a vast, deep liquid realm unparalleled elsewhere in the solar system.

With the Liberation, the French Navy immediately asked Cousteau to train an entire team of divers and form an Undersea Study and Research Group under its auspices. The Navy provided the group with a captured German ship, the *Élie Monnier*, as its diving platform. Under Cousteau's direction, the team removed sunken mines from harbors and continued researching the physiology of diving.

Yet Cousteau, with his restless intellect, yearned to captain a ship of his own, to set off on whatever expeditions inspired him. One evening at a restaurant, the irrepressible Simone struck up a conversation with a nearby table of diners. One of her new acquaintances was none other than Sir Thomas Loël Guinness, scion of the Irish brewing family, who revealed that he'd like to fund oceanographic research. By spring 1950, Guinness was writing a check for a ship. Cousteau had seen a particular vessel docked in a harbor on the island of Malta; instantly he'd known that he wanted her as his own. The boat, having served as a minesweeper for the British Royal Navy, had been built in the United States with a hull not of metal but of Oregon pine to avoid detonating sunken magnetic bombs. By the time Cousteau saw her, she was being used as a ferry. Her new private owner had rechristened her *Calypso* after *The Iliad*'s sea nymph, who in Homer's epic poem haunted the island of Malta and wielded the power to make men immortal. Cousteau kept the name and would maintain the legend.

A farsighted Navy superior arranged for Cousteau to go on leave "in the national interest." In the civilian world, Cousteau was now at last a captain—in French, a *commandant*. He planned to provide scientists with an ocean platform on which they could study, as he later described it to me, "living creatures—not museum specimens impaled by taxidermists." While Guinness had paid for the ship and rented it to Cousteau for a symbolic one franc a year, Cousteau himself was responsible for the costs of maintenance and expeditions. Thus Cousteau the mariner would forever after be burdened with his own kind of albatross, the need to raise money. *Calypso* had to become a working ship—a constantly working ship.

Cousteau managed to attract assignments as well as attention for his *Calypso*. *Life* magazine devoted a full seven pages to her first voyage, undertaken with a team of scientists aboard for a study of corals in the Red Sea. The venerable National Geographic Society, originally founded in 1888 by scientists and explorers, took note. Its Committee for Research and Exploration, which had funded such greats as naturalist William Beebe and Arctic explorer Admiral Robert Peary, initiated a series of grants to Cousteau that would ultimately total a half million dollars. Soon the chief geologist of the British Petroleum Company read about the team and exclaimed to his staff: "These people have imagination!" He funded a *Calypso* expedition to survey for underwater oil in the Persian Gulf. Moreover, for nearly fifteen years *Calypso* would remain the only oceanographic ship to sail under the French flag and so was regularly engaged for quasi-official French scientific missions that allowed the country's scientists to keep pace internationally.

Massive refittings at two shipyards had bestowed *Calypso* with the charms that would one day mesmerize millions: the most modern navigational equipment; an "observation well" built into her prow, down which the team could climb below the waterline to gaze through portholes at dolphins cavorting in the waves; a trapdoor in the galley that led to the ocean, so that divers could enter the sea without being exposed to bitter winds and raging surface waters.

Yet what made *Calypso* unique had nothing to do with portholes and trapdoors and everything to do with the soul of the uniquely unified crew that sailed aboard. Cousteau had a singular gift for making every person, even the youngest member hired to swab the decks, an integral part of a larger effort. They were working for a goal. They were not working for him. They were working with him.

In Jean-Michel Cousteau's biography of his father, *Mon Père: Le Commandant*, mechanic Titi Léandri explained the men's sentiments. "When scientists asked us to do something specific, they took the time to explain what they wanted, why they wanted it . . . We became impassioned: This life had nothing to do with what we'd known in the Navy, where we received orders that we obeyed because we had no choice, where we were divided into very separate classes of men who never met each other. Anyone who explains a goal makes the job more interesting, makes you happy to do anything you're asked. And in that, the *Commandant* was very strong."

Simone Cousteau finally fulfilled her dream of becoming a mariner. She would live forty years on board the ship she always called "my boat," circling the globe ten times. She was said to prefer the company of men and to make any woman who came aboard regret it. The first time I sailed on *Calypso*—also my last—she never spoke a word to me. Her demeanor changed completely some months later when she came to Monaco and saw her husband and me spending twelve-hour days on the manuscript for this book. Shortly thereafter she surprised me by uncharacteristically offering a little sack of tourist memorabilia, posters of Monaco, sample perfumes, even a little coin purse with the word *Monaco* stitched on its side. I knew it was her offering of thanks for the trait she most admired—hard work—the reason why every crewmember respected her and deferred to her. Cousteau proudly described her to me as the real captain of *Calypso*.

Cousteau's good friend Frédéric "Didi" Dumas, whose exploits also enliven this book, joined the first expeditions as chief diver and deputy. Cousteau described him as "half fish"—Didi made more than three thousand dives on the first expedition—and added that because Didi was photogenic "he became our pin-up boy." Classic photos show him trying to poke a shark and demonstrating his experience by catching a stingray by the tail.

Another friend whom Cousteau affectionately described to me was Albert Falco, known as "Bébert." Cousteau often repeated to me the story of the day a prominent manufacturer of diving gear sought him out to tell him, "I have a great gift for you." The gift: an introduction to the twenty-five-year-old Bébert, who would agree to join *Calypso* for one expedition, stay a lifetime, and eventually be named captain.

Even the most celebrated experts who came aboard became devoted members of the extended Cousteau family. Luis Marden had become renowned in his own right not only as chief of *National Geographic*'s prestigious foreign editorial staff but also as the gifted photographer who had revolutionized the magazine's camera work. Marden, on lone dives in dangerous waters, discovered the wreck of Captain Bligh's *Bounty*. After the *Geographic* assigned him to join Cousteau, the two collaborated on pioneering underwater experiments with photographic filters and lighting. "Cousteau had the most original mind I have ever encountered," Marden later recalled. "He was like a Wright brother; and here I was working with him."

The electronic-engineering genius Dr. Harold Edgerton of MIT also played a recurring role on *Calypso*. In the thirties, Edgerton invented the stroboscopic flash for high-speed photography, which flashes twenty million times more quickly than the blink of a human eye. His strobe system for nighttime aerial photography was instrumental in the success of the Normandy invasion. Today we take for granted its innumerable applications: not just the stop-action photography so astonishing in the 1930s but also the copy machine, lights that allow airplanes to navigate and land by night, lights for docking spacecraft, and the illumination of the interior of the eye for medical procedures.

The National Geographic Society had generously endowed Edgerton's trailblazing work. After he had designed the first successful underwater camera, the society arranged a partnership for him with its other stellar grantee of the time, Cousteau. The men on board *Calypso*, for whom the jovial Edgerton had sneaked jars of peanut butter and bright magenta MIT student caps through international customs, soon were calling him "Doc," and Cousteau's two young sons dubbed him "Papa Flash." In the meantime, Edgerton was benefiting from his time on board to create historic sonar devices for seismic profiling of the seafloor. One of his most exhilarating exploits with Cousteau involved photographing the bottom of the Romanche Trench, five and a half miles below the surface. Edgerton created a camera capable of withstanding the bottom's estimated water pressure of eight and a half tons per square inch. He and Cousteau set their camera so that it would begin taking photographs every fifteen seconds from the time of its estimated arrival at the floor of the trench. When the two men hauled the camera back up onto *Calypso*'s decks, they gasped with dismay. The thick glass pane protecting the camera's lens had cracked. They quickly took the apparatus apart. It hadn't leaked. All the same, they had to send the film back to terra firma to be developed before being assured that the camera had captured photos before the lens pane had cracked—taking clear images of the bottom, more than a half mile deeper than the previous record, that prompted even the reticent worldwide community of marine scientists to express their admiration. The ever-amiable Doc responded by presenting Cousteau with the broken lens plate. "Here, Jacques, you keep it," he said. "It will make a wonderful paperweight."

During these years, Cousteau took a step that he never guessed would make all the world his stage. He wrote a book about his adventures on and under the sea, which he titled *The Silent World*. First published in 1953, it went on to sell more than five million copies in twenty-two languages.

While Cousteau had no interest in making a film version of his book, he did nurse a passionate desire to make a feature-length theater film about the undersea world. Accompanied by his friends and crew, he set off on a fifteen-thousand-mile trip across the Mediterranean, the Red

Sea, and the Indian Ocean. At the last minute, his codirector became ill. As a replacement, Cousteau turned to a twenty-three-year-old student who had left one of France's great film schools to join the team. "I was so conquered by the intelligence of this young man that I offered him the position of codirector," Cousteau later said.

That student, Louis Malle, would go on to become one of the great cineastes in French film history. "The four years I spent with Cousteau had an enormous influence on my life," Malle later remembered. "I learned everything about the technique of filmmaking. I had to start my career as an underwater cameraman, as a sound man, as an editor, and as a director, a little bit—a director of fish." Cousteau, Malle, and the team eventually brought a Paris film cutter more than ninety-eight thousand feet of film. They plowed through the footage, created a documentary, and—keeping in mind the popularity of Cousteau's book—titled it *The Silent World*.

In 1956 Cousteau and Malle presented their documentary at the Cannes international film festival. The proceedings glittered with celebrities. François Mitterand officially called the festival to order. Otto Preminger took his seat among the jury. The night that the ultimate prize was to be awarded for best film, the audience included Ingrid Bergman, Brigitte Bardot, Kim Novak, Orson Welles, and Ginger Rogers. And the winners: Jacques-Yves Cousteau and his twenty-three-year-old codirector, Louis Malle. For nearly fifty years, *The Silent World* would remain the sole documentary to have obtained the coveted *Palme d'Or*, the Golden Palm, the highest award of the most prestigious film competition in the world.

Cousteau's eldest son, Jean-Michel, later recalled the klieg-lit evening on the Champs-Élysées when the public at large received its first view of the film at a star-studded Paris premiere. "Everyone who counted in the world of politics and the arts—admirals, generals, presidents, ambassadors—all came to applaud my father." Jean-Michel was seventeen years old at the time. It was only at that moment that it dawned on him, as a typical teenager, that his father was "truly a *sacre bonhomme*"—quite a guy. In the United States, the *New York Times* called the film an "account of oceanographic exploration on and below the surface of the sea [that] is surely the most beautiful and fascinating documentary of its sort ever filmed." Cousteau and Malle went on to win an Academy Award as well.

Yet even all the applause couldn't drown out the siren song that lured Cousteau to distant shores and deeper waters. Before, during, and subsequent to his *Silent World* film success, Cousteau continued exploring. His discovery of a two-thousand-year-old Greek merchant ship, described in chapter 2 of this work, led the then-director of the Scripps Institution of Oceanography to call Cousteau the "founder of undersea archaeology." In later years, some critics objected that Cousteau had failed to recover the wreck according to the academic rules of excavation; more modern excavations have revealed that the remains were not those of one ship, but actually of two, which had sunk some hundred years apart. In fact, there *were* no academic rules of underwater archaeology before Cousteau. Said Patrice Pomey, director of research for France's National Center of Scientific Research: "Cousteau's discovery was spectacular. At that time, no archeologist had ever put his nose underwater. Cousteau's great achievement was to show that the SCUBA could be used to open the field of underwater archeology. He [then] promoted marine archeology to a worldwide public."

On the heels of Cousteau's successes, heads of state paid heed. Rainier III, prince of Monaco, nominated him as director of the principality's world-renowned Oceanographic Museum. The prince's nomination had to be confirmed by Paris's Oceanographic Institute, and the thirty scientists and politicians involved unanimously elected Cousteau to the office. In conjunction with the museum appointment, Cousteau became one of the directors of the Marine Radioactivity Laboratory of the International Atomic Energy Agency (IAEA), a position he would hold for twenty-five years, including the period during which the agency measured fallout from the Soviet and American atmospheric tests of nuclear bombs. He became the secretary general for the International Commission for the Scientific Exploitation of the Mediterranean. He became one of the rare foreigners to be elected to the U.S. National Academy of Sciences. John F. Kennedy invited him to the Rose Garden and personally presented him with the Hubbard Gold Medal for distinction in exploration. The man who continually insisted that he was not a scientist was fast becoming perhaps one of the most scientifically informed, astute, and honored laypeople of his time.

Hand in hand with Cousteau's civic obligations and his explorations came more inventions. Cousteau had participated with Auguste Piccard in the design of the first French bathyscaphe and had personally descended in one of the later incarnations of the craft, an adventure described in chapter 1 of this book. All the same, he was dissatisfied with the machine, a heavy steel gondola made buoyant by large floats. He compared it to a Zeppelin and called it an "elevator without cables." He likewise disdained submarines, which he'd dubbed "cigar engines" for the shape that allowed them to speed ahead in a straight line.

Not for him elevators and cigars. What he wanted, he said, was a "lentil." Cousteau dreamed of a vessel that would allow him to go deeper than he could with an Aqualung dive but with the same mobility. Falco remembers the very lunchtime moment when Cousteau conceived of his solution. "The *Commandant* took two saucers, placed one right side up on the table and the other upside down on top of it. 'There: our submarine!' It was a little quick as an explanation," Falco remembered. Yet within days engineers were working on a "diving saucer." The most inventive touch: Cousteau replaced propellers (deadly below if caught in rocks or abandoned fishing nets) with nozzles that spewed water jets. By adjusting the nozzle angles, the pilot could make hairpin turns around reefs and undersea cliffs. The diving saucer, the first modern maneuverable deep submersible, made the cover of *National Geographic* magazine.

Cousteau himself shrugged off the praise he received for his innovations. "Invention is nothing more than the tool of the explorer," he told me. "If you need something, you make it." All the same it was his inventive experiment with undersea habitats that propelled him into television and unprecedented global celebrity. Cousteau had once imagined that humankind could actually colonize the continental shelf. Thus, when a scientist described his idea for "saturation diving," Cousteau jumped at the chance to join him in a series of experiments the two called "Conshelf," a project described more fully in chapter 8 of this book. Briefly, divers would live in undersea houses so that they could work for extended periods of time below and then, when their work was finished, finally be raised to the surface, sitting comfortably in their habitats during one marathon period of decompression. For the last of his Conshelf experiments, when six men (including Cousteau's son Philippe) lived for three weeks below, Cousteau arranged for live television coverage as the men emerged from their raised and dripping habitat. As they blinked in the spotlights, their faces were broadcast in prime time to seventeen countries.

This extraordinary media coup couldn't have come at a more propitious moment. About this time France launched its own oceanographic ship, and *Calypso* lost its monopoly with the country's scientists. Cousteau was forced to shift his priorities to include significant amounts of time seeking grants and projects to pay for *Calypso*'s upkeep. Around this same moment, however, the National Geographic Society had contracted to produce a series of twenty-seven television specials about exploration. Cousteau's televised broadcast triumph with Conshelf did not go unnoticed, and the society included a documentary about him in its series. The producer was none other than David Wolper, the man now known, among his many successes, for *Roots* and *The Making of the President*. When the two men met, Wolper had already developed a reputation of having intitiated a new, "intimate" style of documentary, in which he showed a personal side of the principal players that allowed viewers to feel they had come to know the characters. Given this style, he and Cousteau were meant for each other.

Wolper set up a meeting with Cousteau and suggested that they join to create an entire series of their own. As he would later say, "Jacques understood everything right away. I said, 'You know, you have those black diving suits that people can hardly see under the water.' And he said, 'Why don't we put a yellow stripe down the arms and the legs?' Almost immediately all diving companies in the world followed him up and put that yellow stripe down the sides of their suits. And he'd come up with the idea just like that! Then he came up with a helmet equipped with an antenna sticking up. The divers could talk to each other and even up to the ship." Cousteau and Wolper won a contract with ABC for a series called *The Undersea World of Jacques Cousteau*, which began to air in 1968. Cousteau would eventually produce more than one hundred television documentaries.

During his decades of exploration and filming, of course, Cousteau kept a keen eye on the changing world around him. His first environmental steps were modest but canny. In 1960, when *Time* magazine devoted its cover to him for having opened the underwater world to sport divers, Cousteau used the moment to declare "with disgust" that he had renounced spearfishing. "We must have a fundamental respect for nature," he said. When he later learned that a train was heading through France to deliver its load of radioactive waste for dumping in the Mediterranean, he relied on such ruses as calling an inexperienced night editor to make sure that his message, which contradicted the political dictates of the day, was published. He managed to mobilize some 11,000 men, women, and children to sit on the train tracks and prevent the dumping. In 1973, he founded the Cousteau Society in the United States. Soon 265,000 paying members became adherents to his powerful environmental philosophy.

It was at this point that I met Jacques-Yves Cousteau. I was living in New York and working for the *Saturday Review*, a magazine that at that time had enjoyed a fifty-year history and was regarded as part of a quartet of thoughtful publications that included *Harper's*, the *Atlantic Monthly*, and the *New Yorker*. The *Saturday Review*'s owner and editor, Norman Cousins, was a tireless proponent of disarmament and a supporter of a strong United Nations. Like Cousteau, he believed that the forbidding challenges of the future required that human beings acquire a global perspective. Cousteau and Cousins decided to join forces. According to their plan, Cousteau would contribute a monthly column to the magazine and the magazine in turn would send issues to members of the Cousteau Society. Cousteau read some of my writing and asked if I would coauthor the column with him.

By this time, given both the brevity of Cousteau's trips to New York and the extent of his business obligations, he could squeeze in editorial work only on off hours. One weekend, in the midst of a soupy New York heat wave, I suggested we meet at the empty *Saturday Review* offices. Although the central air-conditioning would be turned off, an interior conference room had its own thermostat. On the appointed morning, Cousteau and I entered the darkened offices. We reached the conference room, whose walls were occupied by floor-to-ceiling shelves crammed with books. I switched on lights and air-conditioning and spread out our papers on the long conference table with the intention of working on our next column, concerning the intelligence of marine mammals.

At first, we stayed on subject. Cousteau told me about a newborn whale that had bumped into *Calypso*. A female—apparently its mother—had approached and had given the infant a smack with her fin. From there he launched into the tale of his recent discovery of the sunken ship *Britannic*, built by the White Star Line as sister ship to the *Titanic*. After either hitting a mine or having been torpedoed during World War I, the *Britannic* had sunk in 370 feet of water. Cousteau had recently located the wreck. Having no interest in leaving examination of the ruins to robotic submersibles alone, he had undertaken a perilous dive himself—which had allowed him only six minutes to descend, nine minutes to explore, and then had required three hours of decompression. Just at the moment when he swam into the hanging wires and broken planks of the craft, the camera lights went out. Never before had he felt so isolated. In recounting his story to me, he quoted Paul Valéry: "A man alone is in bad company." Whatever the slim relevance of his point to our column about the intelligence of marine mammals—that humankind is not alone as an intelligent being on Earth—we had already left our original subject behind.

We called a coffee shop to order up eggs for a quick lunch. I mentioned some recent shark attacks off the Florida coast. Cousteau gave me a withering look. Never, he said, had he ever known a shark to have "attacked." I looked at him in astonishment. What about the people who'd lost their legs, or their lives, to sharks? Cousteau pointed his fork at my plate: "The sharks are eating. Are you 'attacking' that egg?"

As the day progressed, his tales became ever more intriguing. Violent northeastern winds named *Meltemi* had made it difficult for him to moor *Calypso* in Greek waters near Knossos, even at a modern stone jetty. Where could the ancients have sought safe harbor under such circumstances? On site, he'd pulled out a map and reasoned that they must have moored their boats in a sheltered bay on the south coast of the island of Dia, a few miles off Crete. Just on that hunch, he and his divers had explored the bay and had discovered a vast sunken Minoan harbor. Dia lay in the path of the tsunami unleashed in 1500 B.C. by the eruption of the Santorini volcano, a tsunami that had wiped out cities and civilizations. Could the sunken Dia harbor have inspired the myth of Atlantis?

On and on came Cousteau's ideas: Environmental problems are caused by economic problems; shouldn't we revolutionize the exchange system by replacing the dollar with the kilowatt? And what of the bomb scientists' attempts to patent the nuclear weapon? If the bomb is exchanged, sent from one country against another, who should pay royalties—the sender or the receiver? He told me why he'd chosen Plutarch's *Parallel Lives* as the most important book he'd read. By evening we were still talking and had yet to work on the column. I glanced at the hundreds of books lining the shelves along the walls and knew they contained accounts of all the legendary storytellers, from Scheherazade to Coleridge's ancient mariner. Before me sat their real-life incarnation. For fourteen hours in an empty office building, this mariner had held me spellbound with the tales he'd told. It was a day the likes of which I wondered if I'd ever live again.

Cousteau and I so enjoyed working together that when the contract for the columns ended, he invited me to join the staff of his new Cousteau Society. By that time, I was writing on social issues for the *Saturday Review*—my dream job—and so I declined. But after a few years and a few changes in the ownership, the magazine ceased publishing articles on current events. Cousteau suggested that before I start thinking about searching for a new full-time job, I take a break so that we could coauthor a book. Today I look over his notes to me: "Don't parents hope to give children a better life than they themselves had? Why don't we propose this on a global level? Why don't we try to do for future generations what a parent hopes to do for his child?" He wanted to pass on the "thoughts that had come to mind" after his lifetime of observing nature. He referred to E. F. Schumacher's book *Small Is Beautiful.* "We don't influence future generations only by our genes," Cousteau wrote me. "We also pass on our 'memes'—our ideas." The survival of our species, he said, is linked to the survival of our thoughts.

I accepted his proposal. While the men and women on Cousteau's staff had exciting jobs, accompanying him to far-off lands for expeditions, I far preferred the quiet adventures on which he led me—adventures of the mind. When we were ready to plunge into a new chapter, he'd invite me to his Paris apartment, where I'd take a seat in the living room. Then, for interminable moments, he'd sit in silence, lost in thought. When he finally began to speak, he'd articulate a concise philosophy that clarified one or another complicated issue in the public arena. He'd then send me off to write my own contributions to the reportorial side of our project. Inevitably I would discover that the facts unequivocally supported his contentions.

While Cousteau was originally reluctant to discuss his own experiences in this book, I urged him to include his first-person, eyewitness accounts about events that had shaped his convictions; it seemed to me that the personal experiences that had led him to his beliefs were themselves historic. After all, he was familiar with predators in the corridors of power just as he was with predators at sea. Cousteau routinely expressed distaste for people obsessed with their past achievements. Whenever he completed a film, he never even watched it, so eager was he to proceed to the next. Yet while he disdained living in the past, he never hesitated to learn from it. It was precisely his ability to apply his past experiences to future challenges that make the memories described in this book so meaningful today.

The disinterest that Cousteau showed for his achievements—and the simple, pure pleasure he took in seeking out new ideas and communicating them—carried over into his personal life. I wondered if, despite his celebrity, he was not only modest but in his own charming way insecure. One day as we walked up Third Avenue in New York, we came to a stoplight at an intersection where an elegant woman, dressed in couture clothes and expensive jewelry, was waiting for the light to change. When she recognized Cousteau, she said in Ivy League–accented French that she'd been a longtime admirer. Cousteau gave her a blank look. The light turned green and he marched on. I winced for the woman, left speechless a few steps behind us. I hurried to keep up with Cousteau for about a half-block more when the sidewalk shook with the approach of a huge diesel truck. It screeched to a halt beside us. Up in the cab, a beefy man extended his tattooed arm in a salute. He shouted: "Hey, Jack! Jack Kow-stoo!" Cousteau burst into a grin and began waving both his arms with as much delight as any child greeting a train engineer. In that moment I realized that despite the fact that Cousteau was about as recognizable to Americans as the pope, he had just been intimidated by the elegant woman, not intentionally rude to her.

Cousteau also maintained a sense of humor about his fame. In winter we sometimes took the Paris metro together, with Cousteau wrapped in a furry kind of polyester hooded jacket that he called his "endangered Orlon." In spring we walked. One day on the Champs-Élysées he gave me a panicked look and asked me why people seemed to be laughing at him. I told him that a mime was following us. He chuckled when he realized that tourists from around the world were enjoying Jacques Cousteau in stereo.

I suspect that his good humor energized his other most notable trait, his indefatigability. He made no concessions to age. He was known to have flown across the Atlantic three times in one week. I still have his agenda for a single month when he was seventy-two: He flew from Paris to Washington to testify at a congressional hearing; to New York for meetings; back to Paris; then on to Bogotá for a conference on ecodevelopment. He spent a day flying over the Amazon River basin; he flew to Martinique to board *Calypso*; he attended more meetings in New York; and finally, at month's end, he flew to Montreal to edit a film.

As for my own experience with his indefatigability: Once was enough. Cousteau asked if I'd like to travel with him from Paris to the French Atlantic port of La Rochelle, where the prototype for a new ship, with a wind-propulsion system he and his colleagues had just invented, was being tested. I jumped at the chance, rose at five A.M. in order to meet him at a taxi stand before dawn, and flew with him to La Rochelle. The maritime engineers immediately collared him for meetings, and I spent much of the day walking along the shore. Not until midnight did we board the windship for the test sail. I'd never sailed at night before, and the experience was surreal. In total blackness—with no horizon, no distinction between sea and sky, no perspective—illuminated ocean liners and fishing boats alike seemed to be sailing through space. I stood with Cousteau on the bridge. At some point before dawn, I had an experience unlike any I've had before or after: I fell asleep standing up. I shook myself awake just as Cousteau turned and saw me catch myself. He ordered everyone to retire to the dorms on board. He took care to assign me to a couch and left his own door open so that he could keep an eye out for me, the only woman aboard. After about two hours of sleep, he and I had to leave to catch a plane back to Paris. We landed, caught a cab, and were heading from the airport back to the city when suddenly Cousteau interrupted our conversation and said, "I'm going to sleep for fifteen minutes." He closed his eyes. A moment later his head dropped. He began to snore softly. And then, after exactly fifteen minutes by my watch, he shook to attention and picked up the conversation precisely where he'd left off. After we had entered Paris, Cousteau instructed the driver to stop on the Faubourg St.-Honoré about a mile from where we each had an apartment. I looked around: We were parked in front of the Élysée Palace. Cousteau got out of the cab and told me that he had a meeting with the president of France. I kept the taxi, went home, and fell into bed. It was eleven A.M. I woke up the next morning. I was thirty-three years old. He was seventy-four. When Simone later recounted a nightmare she'd had, I could only laugh. She'd dreamed that Cousteau, having not enough to occupy his time, had taken a job as a night clerk at the Hotel Royalton in New York.

The key to his energy, I'm convinced, was his determination to have fun with everything he did. "The windship is the same as an electric train," he told me. "They're both toys." He loved a story that his father, an inveterate sportsman who swam year-round in open waters and died at the age of ninety-three, once told after having been invited to England for a woman's one-hundred-fifteenth birthday celebration. Journalists had gathered around and asked her, "If you had life to live over again, what would you do?" The lady had replied, "The same thing but more often!" Every time Cousteau related the story he could hardly stop laughing long enough to tell me he felt the same way.

Even as Cousteau and I continued to work privately on this book, his extraordinary public persona continued to take on a life of its own. He became the only French citizen to be voted for seventeen years running to the number-one spot in France's national popularity polls. One year, he received more than eight hundred thousand letters asking that he present himself for the presidency of France. When he declined, commentators understood his decision. They thought that, in the political arena, he perhaps could have saved his soul—but he would have lost the extraordinary independence that he'd already attained by appealing to the global public directly through television.

It would be wrong to suggest in this account of Cousteau's accomplishments that his life was free of doubt. He often expressed the worry that in pursuing his goals, he had shortchanged his family and friends. He and I most certainly had our differences. And tragedy did not spare him. In 1979, his beloved son Philippe, who was also his closest friend and confidant, died in an airplane accident at the age of thirty-eight. In 1990, when Simone died, family members said that Cousteau had lost his star—a perilous tragedy for a sailor. Finally, in 1996, his *Calypso* sank in a freak accident while supposedly safely moored in Singapore Harbor.

He spoke to the press only about his loss of *Calypso*. "I survived three cyclones with her. I was there, gripping the wheel, shouting, 'Hang on! Hang on and get us out of this!' And when you have this kind of relationship, it's not the material things that matter, but the spirit."

Cousteau drew on his spirit. He began raising funds for a second *Calypso*. He married again, to Francine Triplet, who "held his heart." Their two children, Diane and Pierre-Yves, he said, were his "secret to eternal youth." He told *Paris Match* magazine: "Every day of my life is a new point of departure." Soon Cousteau asked Francine to join him at the helm of his international environmental organizations. "I don't feel forced to take up the torch," she told *L'Humanité* magazine. "I do it by passion. Ecology is everyone's business."

In 1992, Cousteau attended the Earth Summit in Rio, where ten thousand experts gathered to discuss global warming and the lack of potable water for the poor. With one hundred heads of state present, photographers were eager for a photo opportunity. Yet after the leaders had gathered onstage, they refused to be photographed unless they were joined by Jacques-Yves Cousteau. Observers agreed that the politicians were not performing a favor for Cousteau—they were hoping for his presence as a favor to them.

Cousteau joined the leaders, smiled as he had his snapshot taken, and then did not hesitate to speak his mind. While he praised the fact that the Rio conference had focused public attention on the environment, he called its conclusions hypocritical. "Rio declares war on poverty, but delegates are offering only one prescription to eradicate it: their so-called sustainable development. What they really are suggesting is economic development, and sustainable economic development is a contradiction in terms. The earth cannot 'sustain' an increased assault on its nonrenewable resources. [The delegates] are saying to the poor: Do as the rich have done! . . . If each individual of the earth's expected population . . . zealously follows that example . . . it will finish off our planet."

All the same, Cousteau never left human beings with a sense of hopelessness. Just about the time of the Rio conference, he was waging what he called his own "holy war" for the future. He had learned by chance that nations had privately agreed to permit exploitation of Antarctica's coal, uranium, and oil. He feared that such development could alter the albedo of the Antarctic continent, could lead to ice melts, and could ultimately raise sea levels by several meters, submerging cities and even entire archipelagos. Cousteau drafted a public petition and obtained three million signatures from people worldwide. The French president, François Mitterand, telephoned him and asked to participate in the effort. The prime minister of Australia came to Paris for a meeting that Cousteau had arranged with the prime minister of France. Because the United States was the strongest proponent of mining, Cousteau spent ten days personally visiting congressmen and senators. Albert Gore focused on the issue, convincing ten senators to sign a resolution calling for the abandonment of mining agreements and proposing that nations negotiate a protocol for the international protection of the Antarctic continent. Cousteau personally pleaded his case with President George Herbert

Walker Bush. Ultimately, the United States joined other nations in signing the Madrid Protocol to protect Antarctica for fifty years—an agreement that many say would probably not have been concluded were it not for Cousteau's efforts.

I cannot offer any definitive memories about a time when Cousteau and I finished our book project, given that even on the day in 1996 when we chose which of our many chapter drafts we'd include in this work, he came up with new ideas for another volume. As we spoke about his thoughts, he talked about writing about his vision of happiness. Happiness, he said, was epitomized by a meal aboard *Calypso*. I said that I'd already gathered as much, having heard stories regarding onboard debates about the origins and endings of the universe, with musicians improvising tangos and Cousteau imitating the Argentine dancers of his youth.

Suddenly, Cousteau leaned back in his chair and gazed at me. "One of my happiest memories dates back some twenty years," he said. "Can you guess what it is?"

I had no idea.

"The day we spent in the empty offices of *Saturday Review*."

I was speechless at the thought that after twenty years the memory I had treasured had been likewise treasured by him. Yet in retrospect, his statement to me characterizes him perfectly. While I, as a young person, had been enthralled by the tales told by an éminence grise, Cousteau, typically, had been enthralled by passing his lessons on to a young person eager to hear them.

In my boxes of documents from those years, I've saved innumerable drawings that Cousteau gave me—his sketches of fish, of "menfish," of undersea grottos and undersea habitats. But the one page that means the most is the simple two-word note he left on my desk on a Friday night. He wrote: "*Bon Courage*." Have courage. I keep it now, not as a memory of how he hoped I'd proceed with work that weekend. I keep it as his message for how he wanted me—and all of us—to proceed on the path ahead without him.

CHAPTER ONE

THE DRIVE TO EXPLORE

W HEN I was six years old, locomotives fascinated me. My parents had rented a flat for the summer in a small house by the sea in the western outskirts of Cannes. A railway traced the shoreline of the French Riviera; trains would materialize out of the distance, speed along the Mediterranean, and then thunder past on tracks that wound just behind our house. During those long summer days, I spent hours on the beach, spellbound, watching various kinds of trains hurtle by. The hissing black steam engines filled me with wonder; to a little boy, they were enormous dragons, roaring and stampeding on their way off to elsewhere.

In the distant sky, a wreath of smoke wafted above the spot on the horizon where the locomotives dropped out of sight. Could it be that way out there, under the faraway smoke plume, scores of the great iron beasts converged, all fuming and snorting before they once again lunged to a start, pulling their massive caravans toward the ends of the earth?

One day, the riddles of this magical smoke ring proved too much for me. I abandoned my toys on the beach. I headed for the tracks. I stepped between the rails and began to run, following the gleaming parallel lines that pointed to my dreams. Before long I lost my breath and had to slow down, but an intense yearning had overtaken me: I could not stop; I could not turn back. Panting, tumbling on the ballast pebbles, I pressed on and on, interrupting my steps only when the ground shook as the fearsome convoys approached from behind and I had to scurry out of the way. I had never seen the tremendous engines so close. I could feel my heart pound with the beat of the rails.

By evening the smoke wreath, now outlined in the twilight, seemed just as far away as ever. Confused, exhausted, I lay down to rest for a minute on the gravel. I was drifting off to sleep when some policemen found me and, despite my furious protests, took me all the way back home. When they delivered me to my anxious mother and father, I didn't try to offer my parents an explanation. I knew from the tears in their eyes that they would not understand how I had ached to reach that horizon, why I had been unable to resist running after it.

I did not understand, myself.

Even today I could not hope to explain the drive to explore, although I have readily recognized its telltale signs of unrest in the eyes of explorers I have been privileged to meet: The pioneer speleologist Norbert Casteret, so eager to crawl deeper into the inner mazes of the earth that his colleagues nicknamed him *Le Cavernicole*, after insects that can live only in the total darkness of recessed caves. Haroun Tazieff, the intrepid volcano explorer who—standing on the edges of flaming craters, under pelting rains of red-hot lava—sniffs the air and casually remarks, "It smells of sulfur." Admiral Richard Byrd, who with an enigmatic smile once told me, "As long as you haven't been to the polar caps, you won't understand why I have to return there." And Théodore Monod, director of the French Institute of Black Africa in Dakar, whom I met when we both participated in the first of Auguste Piccard's bathyscaphe dives. When I later rigged *Calypso* to sail the Red Sea, I invited Monod to come along, and at first he eagerly accepted. Then he asked which coasts we would explore, and I answered, "The Farasan reefs off Lith in Saudi Arabia." Even as Monod declined my invitation, his mind already seemed refocused on a different goal. His own unending quest drove him to seek the secrets of Africa. As he explained it to me: "One continent per life."

In fact, the drive to explore has inflamed the human soul for centuries. I doubt that anyone so consumed has ever understood exactly why. Daniel Boorstin, in his celebrated history of exploration, *The Discoverers*, noted that early wayfaring merchants and fortune seekers, unlike explorers, set off with their eyes on specific prizes, abundantly clear in the names they bestowed on newfound shores—the Ivory Coast, the Gold Coast, the Grain Coast, even the Slave Coast. But what motivated explorers? What inspired Magellan, battered by South America's strange williwaw winds, to hold to his course through an unknown strait with no guarantee that it would lead to an untraversed sea? What makes adult and child alike feel so desperate at the prospect of abandoning their advance along shining rails, across shining seas, that lead beyond the boundaries of their familiar world? What inspires an explorer to undertake a voyage with no destination, to search with no objective, to travel with no itinerary other than the uncharted, the unfathomed, the unexpected?

To enlarge the human perspective, to build on knowledge for future generations, to identify dangers, and to chart the course to a better world: If these are the goals of the explorer, then everyone—voyager, scientist and citizen, parent and child—is engaged in humanity's momentous expedition. In that sense it may be interesting to ponder why some of us feel so compelled to set out on literal explorations. Perhaps some common trait links the world's explorers and at least partially explains why certain men cannot resist the lure of knowing the unknown. Historians and sociologists may one day identify it. As an explorer, I cannot dissect the drive to explore; I can only describe it, by telling my own tale.

The urge to explore seemed so basic to me when I was little that I wondered if nature endowed all forms of life—animals, even plants— with an exploration instinct. I noticed that the wisteria shoots creeping up our garden wall turned away from shade and instead infallibly searched out the best direction for growth. I listened raptly as my *oncle* Joël, who kept hives near his vineyard to supply the family with honey, described the sallies of scout bees that explored our entire neighborhood.

Human beings, too, seem to be instilled with an instinctive drive to explore, at least at birth; even the most fainthearted adult remembers the thrill of a childhood expedition beyond a forbidden door or into the darkness. When I was growing up in the little village of Saint-André-de-Cubzac—which, in those days, had no electricity—the children in my family eagerly awaited the adventures that came with the waning light. After dinner, the grown-ups would gather around our huge fire-place and immerse themselves in conversations that soon became incomprehensible to us, the five bedeviled children of three families. Our ageless maid, Seconde, invariably wrapped in black from head to toe, would call the young ones to the stairs and hand each of us a small kerosene lamp—*une lampe Pigeon*. Sometimes, instead of going to our bedrooms, we would tiptoe on up another flight and sneak into the attic. There we would startle one another by springing out of hiding places behind dusty coffers and dented trunks. In the eerie darkness of a far corner hung a skeleton, deemed useless for anatomy classes when it lost a few bones but still possessed of more than enough magic to capture our young imaginations. As we moved our lamps, its shadow haunted the attic; it slipped between valises and suddenly loomed up over us, across the ceiling. When I was last in line to leave, I sometimes would dare myself to turn and linger alone for a minute. I would watch the ghostly silhouette dance gleefully among the spiders on the wall and, with a shiver, wish that somehow I could enter its world of mysteries and dreams.

Soon enough I discovered that the real world—my world—offered far more provocative mysteries than the human mind could imagine. My father arranged for me, at age ten, to spend two weeks at an American summer camp on the shore of Lake Harvey, in Vermont. A huge, forbidding German man served as camp counselor. My fear of riding horses exasperated him; to punish me, he commanded that I clear out the reeds entangled under and around the diving platform in the lake. What a punishment! For two weeks I spent each day diving, each night restlessly waiting to dive again. As I entered my teens, I eagerly progressed from simple diving to simple diving experiments. I'd read sagas of early American adventurers who threw Indians off their trail by submerging themselves in river waters and breathing through reeds. The French region of Alsace-Lorraine wasn't the American frontier, but good enough: My family spent a summer there in a house with access to a swimming pool. I tried to sit on the bottom, breathing through a tube I'd rigged to reach the surface. My experiment's failure gave me my first exhilarating success: I had discovered for myself that under a pool's worth of water pressure, human lungs cannot draw in thin surface air.

After a few more years, pools and lakes alone no longer quelled that familiar nagging need to see more, discover more, learn more. As a young adult, I yearned to sail the seven seas, to travel to exotic ports, to the South Pacific, to the fabled Orient. The Navy offered an immediate means to my ends, and so I enrolled in the *École Navale*. Yet each time my ship docked, I ached to move on. Even the lure of Shanghai vanished as soon as I got there; centuries ago, after all, other Western voyagers to the Orient had claimed for themselves the breathtaking prize offered by every unknown land—the first vision of it.

Like the other officers, I focused my interests ahead, on the ports to which our ship sailed. I gave no thought to the underwater world below our keel. Then, one day in Indochina, we began a survey to map the Bay of Cam Ranh. A native rowed us from one spot to another as we recorded our measurements. Shortly past noon, just as the heat of the day peaked, the rower lifted his oars out of the water, signaled us to remain silent, slid over the side of the boat, and slipped soundlessly into the water. With no mask on his face and nothing in hand, he flipped heels over head and disappeared. A moment passed. A splash; he surfaced; up shot his hand, grasping a few fish! Our interpreter patiently explained to us wide-eyed Westerners that sometime around noon fish take siestas. Our oarsman liked to pluck them from the water while they napped. This improbable notion—fish take siestas?!—amused me but it also transfixed me. I glanced at the surface of the bay, the crests that curled at the prow of our boat—here at last were the gateways to the last remaining unexplored expanse of earth, to the undiscovered world I'd always dreamed of entering, whose endless wonders, I sensed even then, would never quench my drive to explore but forever fuel it.

The silent world, of course, offered visions and hopes but it did not necessarily offer a job. What factors triggered me and the other explorers I've known to launch actual careers in exploration? Jacques Monod and Democritus believed that "chance and necessity" turn the wheels of evolution. Some might say that chance events and the necessary response to them set explorations rolling as well.

Certainly my wife Simone and I did not believe, as we put our four- and one-year-old sons to bed on November 27, 1942, that destiny was poised to change our lives. Even the war did not distract us from our immediate goal: We planned to leave Marseilles the next morning for Lisbon, where I would begin my commission as an assistant naval attaché. I had been participating in the Resistance movement; the Navy had assigned me to Portugal so that I could serve as an intelligence link between London and contacts I had developed in Marseilles.

That night Simone and I were exhausted after a day twice pierced by shrieking air-raid sirens. Nonetheless we forced ourselves to ignore our need for sleep and hurried to stuff all our belongings into trunks and suitcases. As we packed, we listened absently to a tinny tune on the radio.

The music stopped. An announcement crackled over the wire: To prevent its ships from falling into the hands of the invading Germans, the French Navy had scuttled its fleet in Toulon.

I switched off the static of the radio. Simone and I sat in silence. Tears streamed down our faces. We had lost our fleet. Our pride. Our hope. Our last remaining shred of French independence.

The next day my assignment to Lisbon was canceled. I would remain in Marseilles, continuing my underground activities—and using my professional diver's card as a cover. Thus I became an undersea explorer.

Countless others have also become explorers after meeting up with unanticipated events—tragedy, rivalry, bankruptcy. Nonetheless I am not persuaded that happenstance and luck are prerequisites for an exploration career. If the Navy had not scuttled the fleet, if it had not canceled my Lisbon assignment, would I have simply disregarded my lifelong urge to explore? Or would I have jumped at the chance to explore the Atlantic waters off Portugal? In fact, as soon as I had received my Lisbon orders, I had begun to savor that very possibility. True, no one can absolutely control the direction of his life; but each person can certainly influence it. The armchair explorers who complain that they never got their "one lucky shot" were never really infected by the incurable drive to explore. Those who have the bug—go.

Every explorer I have met has been driven—not coincidentally but quintessentially—by curiosity, by a single-minded, insatiable, and even jubilant need to know. Given, the wisteria shoots that seek the sun, the scout bees that report the best new nesting sites back to their hive—all represent life forms that, in nature's complexity, somehow end by flourishing in their best environment. Yet *Homo sapiens* stands apart because of his questing intellect. The French philosopher Montaigne celebrated the advent of humanity's great intellectual gift by quoting Ovid's line: "While other animals face down to earth, God raised man's face to heaven and bade him contemplate the stars."

This capacity for curiosity explains the most bewildering aspects of the exploration drive. To those who wonder exactly what practical goal the explorer takes, the answer is simple: none. Doc Edgerton, who joined us many times on *Calypso*, once said it precisely. He lowered his camera directly into the sea's Deep Scattering Layer to help us explain this mystifying line that had been appearing on our ecographs, rising then descending with night and day. He sank his camera more than five miles down into the Romanche Trench, and eventually, he sought out the secrets of the deep seas in more than ten thousand photographs. One day a young man, nonplussed by Edgerton's insatiability, questioned him. "Let's grant that someday you'll design a camera that won't collapse under the pressure of extreme depths. Let's grant, further, that you find a way to lower it safely into position and start shooting. What do you expect to find? Giant squid? Buried cities? Sea serpents?" Edgerton looked at him blankly and said, "If I knew what we'd find, I wouldn't bother to find it."

The bureaucrats who can't grasp the fact that explorers are propelled by simple curiosity—not ambition for a particular discovery—exasperated the great biochemist Albert Szent-Györgyi. In a remarkable letter to *Science* magazine in the sixties, Szent-Györgyi wrote that the basic scientist, a true explorer, "knows only the direction in which he wants to go out into the unknown; he has no idea what he is going to find there and how he is going to find it." How, then, to write a proposal asking for research funding from goal-oriented reviewers? "I always tried to live up to Leo Szilard's commandment: Don't lie if you don't have to. I had to. I filled up pages with words and plans I knew I would not follow. When I go home from my laboratory in the late afternoon, I often do not know what I am going to do the next day . . . How could I tell, then, what I would do a year hence?" The startling discoveries Szent-Györgyi made by following his curiosity—for which he'd been awarded not one but two Nobels—in fact so contradicted all prior assumptions that even if he had been able to predict them in a grant application, the popes of the field would have rejected him out of hand.

This roving curiosity sooner or later compels an explorer to take only his quest seriously, not himself. Certainly my friends and I reveled in the game of exploration. Philippe Tailliez, my colleague even before our first trials of the Aqualung, sometimes could not even contain his joy. One day, Tailliez and I dived off Saint Raphael, eager to examine in detail a madrepore-covered rock we'd seen. We had to descend into deep waters to reach the spot, but Tailliez and I had long since become accustomed to each other's relative safety and distress, often warning each other by shouting into our mouthpieces; water distorts words too much for complicated conversation but carries loud noises well. When we reached the bottom, we found that a strong current surged across the sands, but our interest in the rock ran stronger. We were absorbed in our findings when Tailliez suddenly shouted—on and on, a long sequence of incomprehensible yelps. I signaled him to repeat. With great effort he started blurting sounds again, which, mingling with his bubbles, turned to gargles by the time the water carried them to me. I peered into his face mask. His eyes bugged; his cheeks swelled; his lips and skin cast greenish shadows. Forgetting that the depths always rendered my own face as hideous as his, I became alarmed. Now he hovered so close to me that our masks touched; his pop eyes stared into mine as once again he bellowed senselessly. I grabbed his shoulder and urgently signaled a return above. As soon as we surfaced, I asked him what was wrong. Answered an incredulous Tailliez: "With me? Nothing. I was singing you 'La Marseillaise.'"

Fun and curiosity together become addictive, making an explorer not just exuberant in his need to know but also relentless. Our expedition to the island of Abu Latt in the Red Sea gave me my first realization of just how tenacious a curiosity-driven explorer can be. Even those who have seen Abu Latt must wonder if they dreamed it; the island's beaches, made not of sand but of pulverized coral, are lapped by a shallow lagoon, which in turn is rimmed by a barrier reef that drops off into the depths of the Red Sea. For our first Aqualung dive in the area, we kicked lazily across the lagoon to the reef, then gazed down the vertiginous coral cliff. We began to descend along the jagged, multicolored wall. Frédéric Dumas, a diving ally since before my first Aqualung days, spotted a small opening in the coral and stopped, enthralled. He poked and probed, soon completely absorbed by the pure fossil coral. I looked above: Black silhouettes of sharks made arcs through silver light from the surface. I looked below: Sharks glided across the sandy bottom shoal. I looked at Dumas, who remained engrossed. My heart stopped: Out of the distance appeared an enormous *Carcharhinus* shark. It shot toward Dumas's back. He continued examining the coral with the shark swerving, circling back, sniffing at his ankles. I shouted into my mouthpiece, to no avail. Dumas continued scrutinizing, oblivious to the thirteen-foot shark swooping in figure eights back and forth behind him. He didn't even notice when the creature turned and, in a flash, disappeared into the depths.

The image of men so preoccupied with coral specimens that they ignore wild carnivores made me wonder if the drive to explore had driven us mad. Yet however extreme our own displays of single-minded determination, they couldn't compare with the resolute drive of the circumpolar navigator David Lewis. *Calypso* was sailing the South Polar oceans, seeking shelter from the most punishing waters of the world, around Hope Bay. At four o'clock one morning, a clattering bang against the hull woke me. I jumped out of my bunk toward the porthole. A strange, nearly demolished little yacht, mast broken, was bobbing in the ice water and smacking against our ship. I ventured outside. To my astonishment, out of the wreck emerged the oddest storybook character, beard tangled and dripping, hair grown down to his shoulders. I squinted through the darkness and drew in a breath. He was in bad shape. His hands were swollen. His fingernails were black. He could barely walk; he spoke only with difficulty. We tied his wreck up to *Calypso*, helped him aboard, and I walked him, hobbling, to the mess, where I eased him down at the table and put hot food in front of him. All ships in the area—including *Calypso*—had been alerted to look for a solitary mariner, missing for two months. I telefaxed NASA to say we had him aboard. Dr. Lewis had sailed twenty-five hundred miles from Australia in his little boat, the *Ice Bird*, when it had capsized and broken its aluminum mast. He had improvised a makeshift mast from an emergency short boom and somehow survived for eight weeks in freezing sea waters and in his unchanged soaking clothes and ice-water-filled boots. As soon as he was warm I offered to transmit, by satellite, any handwritten message he wanted to send his ten- and twelve-year-old daughters. After reassuring his girls and sending them his love, Dr. Lewis—clinging to life after a harrowing ordeal amid raging seas and heaving icebergs—wrote of his plans: "So the boat is a bit broken but I will manage." He repaired the *Ice Bird* and set off for the Cape of Good Hope.

As often as I shake my head in amazement at Dr. Lewis's determination to sail farther, I never understood people's bewilderment at my own zeal to go deeper. The odds we encountered since the early days—our convulsions caused by breathing pure oxygen more than thirty-three feet down; the threat of embolism and onslaught of sharks; the death of my fellow test diver Maurice Fargues, at 397 feet, from nitrogen narcosis—all aggrieved us, but did not defeat us. Nor did the triumph of reaching a new depth ever silence the siren call from waters deeper still.

And so, in 1953, during the French Navy's testing of its FNRS3 "Deep Boat," for which I served as technical adviser, I seized the opportunity to descend into realms forbidden to the Aqualung diver. As I climbed into the bathyscaphe, Lieutenant Commander Georges S. Houot, its captain, pointed to the distant peaks of Mt. Coudon, rising majestically 2,303 feet above Toulon Harbor. We would descend into the sea, he said, more than double the distance that its crest rose into the sky. I settled into the craft and looked out the porthole, where we'd affixed one of Doc Edgerton's electronic flash units. If it could not withstand the pressure of those depths, if it imploded, it could blast open our hull under almost a mile of water. Yet as the craft dropped through the surface and sailed into darkness, all the deadly possibilities faded before the dazzling realities. Here, at 400 feet: the billions of tiny organisms that constitute the mysterious Deep Scattering Layer, which Doc Edgerton had so often photographed for us. There, at 1,000 feet: bizarre fish with bulbous eyes, squid that brightened the sea's eternal night with glowing clouds of luminous ink.

The siren who'd been calling me did not disappoint, but neither did she let me be. Her banquet did not satisfy me but only made me long to feast my eyes again. And so, on July 24, 1954, undeterred by the cast on my broken foot—would Antarctica's determined mariner, Dr. Lewis, have laughed?—I limped into a bathyscaphe again. The spectacle continued. What looked like unearthly snow fell upward—tiny creatures, hanging in the water, their apparent ascent an illusion created by our downward motion. Finally, barely discernible in the edges of our spotlight, I made out an amorphous shape, the color of mud. Already the bottom?! I asked Houot to slow for landing. As our dangling guide chain sank into the slime, the view from the porthole astonished me. We had anchored, not on the bottom, but on an unsteady shelf of sliding mud that bulged from the vertical continental cliff. Though our depth gauge indicated that we had in fact descended 4,920 feet, the cliff dropped off, precipitously, into a yet deeper canyon below.

Intent on reaching the bottom, we prepared to drive the craft off the ledge. Houot started the engines. Something tugged. Our guide chain had stuck in the mud. The craft lurched against the resistance, and then the chain suddenly broke loose, ripping off a giant block of mud that toppled off the ledge and plummeted into the abyss. As it plunged, it tumbled repeatedly into the mud cliff, knocking loose more blocks. A brownish yellow silty cloud roiled up from the bottom.

"Houot, we've started an avalanche!"

We hovered in the bathyscaphe, hoping that we could salvage the dive for photography by waiting for the mud to settle. Twenty minutes passed. Still the thick yellowish clouds rose and blossomed up from below. Could we find clear water if we navigated by compass, driving right through the clouds across the canyon to the other side?

We steered the bathyscaphe out over the abyss, clouds billowing below us, occasionally enveloping us. From time to time, between clouds, suspended debris in the clear water seemed to speed by our porthole as we passed. Our motors whirred. More than enough time elapsed to reach the opposite canyon wall.

Then came a sickening realization. Particles had ceased to whiz past our window. The porthole was blocked, covered by a thick layer of yellow-brown mud. Neither Houot nor I said what both of us wondered. Had we run blindly into the opposite wall, toppled it as well, and, now at 5,250 feet below the surface of the sea, been buried alive?

Or could it be that nothing more than one more cloud had swelled up around us? Perhaps the muds would settle. We waited an hour.

Nothing.

We decided to try to surface. Houot released as much ballast as was necessary to send us sailing upward.

The bathyscaphe remained motionless.

To conserve oxygen, we spoke little; but our thoughts raced. An idea: During the eternal hour we'd waited in the frigid, sunless depths, had our gasoline cooled and become heavier? Houot released more ballast.

Up we soared.

Thus I learned the single essential to qualify a dive as successful: a return.

This lighthearted proposal in fact represents the weighty truth of all exploration. Voyagers wander; explorers return, with tales of their travels that enrich their homelands and push back the human horizon even for those who stayed behind. As Daniel Boorstin wrote, vagrants from ancient Rome, traveling on wind-driven vessels, seem to have left coins in Venezuela; in pre-Columbian times, Chinese or Japanese junks may have been driven off course and swept all the way to the shores of America. But the travelers never reached home. "These acts and accidents spoke only to the wind," Boorstin wrote. Feedback is the explorer's commodity. The ships of the great explorers were those capable of coming home, those that delivered their precious cargo—news.

Marco Polo, taken prisoner in Genoa during its Venetian war, dictated his accounts of the great Kublai Khan to his cell mate. Darwin took his own copious notes with pen and ink. The twentieth century gifted humanity with a universal language—film. Thus it seemed natural to us to record the wonders we'd discovered, exploring with *Calypso*, for cinema. To produce science documentaries and science journalism was never our ultimate goal. We simply wanted to share nature's hidden secrets with those for whom they would otherwise remain inaccessible. We never attempted to decipher the meaning of life; we wanted only to testify to the miracle of life.

And so while scholars toiled at their desks, we observed nature's vast expanses. As scientists alerted the world to a greenhouse effect that warms the atmosphere, to the prospect of melting glaciers that could flood coastal cities, we witnessed the planet's own testimony to its changing sea levels. In the Mediterranean, east of Carthaginia in Spain, we dived along the side of an undersea mountain and found an inundated cave, carved long ago by lapping surface waters. On a ledge at its mouth rested a neatly piled mount of oyster shells—left, most likely, by the humans who once lingered on this fossil beach, now 558 feet below the surface of the sea.

As anthropologists counseled against anthropomorphism, the mistake of attributing human qualities to animals, we discovered we had much to learn by the inverse, observing animal characteristics to learn about humans. When elephant seals overpopulated the island of Guadeloupe, the animals formed a veritable sex and milk bank. Bulls coupled indiscriminately with available females; mothers nursed whatever pup approached a teat. In scarcity, the seals had obeyed strict societal laws. In opulence, all order disintegrated.

As sociologists lamented the breakdown of the human family, I watched from a helicopter as a male orca risked his life to save his kin. The *Calypso* team made a valiant attempt to film the entire orca gam together. But when our men approached the whales, the lead male made an open display of himself and successfully lured the ship away as his family members dived and disappeared. With our hydrophone, our team listened as he whistled, perhaps directing his little tribe to safety. Then he too dived and vanished. From the helicopter above I searched for him and finally spotted him, reunited with the other whales miles away. All day, our team tried to film the majestic group; and all day the lead orca repeated his ruse, confounding my men before he plunged below the surface and unerringly found his way to his distant family.

As physiologists emphasized the human need for fellowship, we witnessed the significance of camaraderie among dolphins. We've often been mesmerized by the spectacle of a thousand dolphins, jumping and leaping and spinning in exuberance, swimming in choreographic unison, their silky bodies sliding against one another, their flippers often touching. We've seen dolphins lift their young and even their injured companions to the surface, supporting them as they take in air through their blowholes.

Fellowship seems to be as vital to dolphins as oxygen. For years I never saw a dolphin alone. Then one day, while crossing the Mediterranean on our way from Corsica to Sicily, we encountered a single dolphin, floating nearly motionless on a dead calm sea. I asked that *Calypso* approach him, slowly. The dolphin did not react; he merely kept his eyes above water, watching us. We reached out to touch him; he did not flinch. Thinking that he must be gravely ill, the ship's doctor asked us to hoist him aboard and put him in our tank for examination and whatever aid we could offer. The doctor found the creature's breathing to be absolutely normal, as well as his temperature and pulse.

The next morning the dolphin was dead. An autopsy showed every organ to be sound. The doctor's diagnosis: death from loneliness after probable exile from his pod. The creature's body was not broken; his spirit was.

As industrialists, technologists, and environmentalists debate the relative risks and benefits of progress, we acquired a film record of the changes that one species, the human species, has imposed on all the others. In 1946 fishermen wove extraordinary stories for my friends and me, telling us about a rock called Le Veyron, around which sea life swarmed. We boated out to the area, climbed overboard into the sea, broke surface, and found ourselves amid an undersea paradise. The lush terrain transcended even the marine jungles we had seen; it spread out before us with all the luxuriance of an impeccably manicured English garden. The seascape itself was no ordinary rock, but a complex of caves, stone archways, and narrow passages that wandered between steeply sloping walls. Every exposed surface was carpeted with colorful carcareous algae, green bushes strewn with checkered patches of sea urchins, and clusters of potato-shaped ascidians. Screens of red and yellow gorgonia and brittle pale-pink bryozoans sprang from scattered rocks. The labyrinth of stone compartments provided even better living quarters than a shipwreck. Octopuses crawled in crevices, corbs thronged in the niches, rays lolled in the shade of stone pillars. Huge groupers, sixty-pounders, shoved their snouts out the entrances of their lairs and stared curiously at us with their globulous eyes. Our cameras whirred, our heads turned, our eyes drank in the intoxicating sights.

About thirty years later I returned to Le Veyron with my *Calypso* team. We dived to the same depth, to the same caves, at the same time of year. The grotto was empty. Not one single fish lived among the rocks. The verdant gardens were gone. Straggling sponges and a few clumps of grass were studded here and there like scattered tombstones. Human beings had polluted the seawater and mechanically destroyed the nearby coast; all life had paid this price.

Often, in airports, on sidewalks, at restaurants, children and adults alike stop me to ask about barracuda and sharks; killer whales; the deadly sorcery of the Bermuda Triangle; the Loch Ness Monster. When I saw Le Veyron, I believed that the sea's most monstrous force doesn't live in Loch Ness. It lives in us.

Shortly after I left Le Veyron, the governor of California invited me to speak at a prayer breakfast. My role as an explorer, until that time, had been simply to witness the miracles of creation and to share them. A prayer breakfast hardly seemed the place to curse the darkness.

As I stepped up to the podium and faced the members of the audience, I realized that, despite the diversity of individual beliefs held by the many people present, we all shared a faith—a faith in humanity as the only species of Earth with a capacity for reverence for life. Given humanity's singular gift—the intellectual capability for protecting life—it seemed to me unworthy to decry a problem rather than to solve it, to damn progress, technology, industry, every human advance intended, after all, to elevate the human condition.

It was then that my life of exploration produced a personal discovery: Logical absurdities, I realized, were preventing the human community from what I consider reasonable utopias.

I'd heard industrialists contend that they cannot afford the luxury of expensive toxic waste disposal—a superficially logical statement, but nonetheless absurd in its suggestion that we will enhance the economy by jettisoning deadly poisons into our rivers, lakes, and soils. And why not save the community the burdensome costs of repairing damage done, by raising the prices of the products in question in order to pay for proper disposal of the wastes generated during manufacture? Reasonable, but utopian.

I thought of other social and environmental problems I'd encountered, and the list grew longer. The logical absurdity of fishermen who respond to declining fish populations by doubling their efforts for a catch, thereby wiping out the stock and proceeding from loss this year to bankruptcy the next. The logical absurdity of believing we must apply every single scientific finding we discover, of thinking that progress requires subordinating human interests to new technologies rather than using new technologies in the human interest. The logical absurdity of military officers who propose offsetting the dangers of their neighbor's towering plutonium and bomb stocks by augmenting their own stores. The logical absurdity of trying to bolster the world economy by institutionalizing the mass market, enriching the rich and impoverishing the poor. The logical absurdity of leaders' professing to support human rights while they ignore the rights of future generations.

Many scholars have contended for the past few decades that the great age of exploration has ended. Casteret and Tazieff, Hillary, Piccard and Amundsen, the Vikings, back even to Pytheas—all are adjudged to have been born, as I was, at a special time, a time when explorers could be adventurers, when big discoveries could be made by a lone, determined person or by a small team. Now that the earth is mostly mapped, and space and scientific exploration require tools like billion-dollar rockets and atom smashers, no mere individual, they say, can hope to become an explorer.

As I looked over the expectant faces of those gathered for the governor's prayer breakfast, I realized that the age of exploration hasn't ended; the role of the explorer has simply changed. These people were searching too: They were searching for hope. They were not defeated by my somber litany. They knew that the solution lies in ceasing to regard tragedy as inevitable and in finding a way to make utopia attainable.

We, the new explorers for the modern age, must rely to the fullest on the traditional explorer's greatest asset: the human intellect. Unlike any other species on Earth, human beings have the ability to look back in time and to learn. We know the strengths of the great ancient civilizations as well as the catastrophic mistakes that crumbled empires. It was the traditional explorer who blazed a trail to this past. The new explorers—all of us—must use that trail to find our way to the future, to chart our course, and our children's course, into the unknowns ahead.

The past age of the lone explorer, Boorstin wrote, opened at a time when medieval cartographers did not place "North" at the top of their maps. At the upper margins of their parchments they drew instead a mysterious spot they imagined in the east: paradise, which no living human had ever entered. Monks, looking for this garden "where Earth joins the sky," became the first explorers; next, pilgrims wandered in search of the holy place; then crusaders forged into new lands in the name of God; and finally, missionaries infiltrated farther into Earth's unknown corridors—all of them pushing back the boundaries of their world in quest of paradise.

On one of my own voyages, I think I found it.

Calypso was exploring the waters of the Aleutian Islands, in Alaska. We sailed into an unreal world—the sky, the ice, the water, all brushed by the indescribable light that makes dusk indistinguishable from dawn during the season of the midnight sun. After I dropped anchor off the island of Unalaska, our chief diver, Raymond Coll, climbed into our diving saucer and took it down for a search under the surface of the sea. Raymond, who stayed in contact with me by sonar telephone, described walruses and whales, vast pink fields of krill—icy waters that brimmed with life.

Above, on board *Calypso*, the team was uncharacteristically preoccupied with another endeavor altogether. NASA had just landed a module on the moon, and the entire *Calypso* crew crowded onto the fantail to listen raptly to the radio, awaiting the announcement that a human being, Neil Armstrong, was about to step onto the lunar surface.

Excitement electrified the atmosphere on board. We barely breathed. And then, Armstrong's voice. A cheer rose from the deck and echoed in the crystal air. I switched on my sonar link with Coll. "Raymond! Historic news! Armstrong is walking on the moon!"

I felt a proud sense of fraternity that day with the NASA explorers who were reaching into the distant corners of space even as our little team explored the depths of the sea. Yet I also felt a sense of privilege that had to do not with the astronauts, but with the wet, blue Earth on which they were gazing. Out in space, a human being stepped through dry lunar dust. But Raymond sailed through a myriad of life, undersea on the solar system's only planet awash with liquid water.

Several years later, the astronaut Alan Bean spoke to me about his own space voyage. From the moon, he watched Earth spin through cycles of night and day. When the dark side of Earth faced him, our planet appeared to be encircled by a girdle of sparks. It took him quite some time to realize that these tiny glimmers were flashes of lightning from the quasi-permanent thunderstorms that ring the equator. He marveled: "The Earth looked like a scintillating jewel."

And that it is—the very reason why searching for ways to safeguard Earth is such a necessity and such a joy. After all, the road to paradise—as the Spanish proverb goes—is paradise.

CHAPTER TWO

PERSONAL RISK

My introduction to the deadliness of unnecessary risk caught me off guard. During the summer of 1952, the possibility that tragedy would befall my team had not even entered my mind; I was completely absorbed in our most recent discovery. A diver friend had boasted to us about the number of lobsters he had caught near a sunken heap of "old jars" about ten miles off Marseilles. The words *old jars* resounded in our ears. We knew that the remains of ancient sailing vessels, still laden with the elegant flasks in which they carried cargo, lay scattered across the Mediterranean's floor. Could these "old jars" be—? We raced to the spot in a launch, dragging along an antiquities expert. As soon as I had retrieved a few pieces from the bottom and surfaced, the professor grabbed the dripping pottery out of my hands and began exclaiming. The jars were from ancient Greece; they were more than two thousand years old.

I could barely contain my excitement as I clambered into the boat and described the extraordinary vision on which I'd laid eyes 140 feet below. Hundreds of amphorae lay buried up to their necks, their exposed mouths neatly aligned in rows that outlined the shape of a now vanished vessel, with pieces of dinnerware, which had scattered as the ghost ship sank, still strewn nearby. Surely the site marked an ancient merchant ship—a beacon from antiquity—undoubtedly older than any seagoing vessel that had ever been excavated.

We eagerly met the formidable demands of a major recovery operation—toiling in dangerously deep water, hauling down a suction pipe to remove tons of sediment that had settled over the invaluable pieces, transporting artifacts to the surface for scholars awaiting at a Marseilles museum. We worked incessantly all that summer, salvaging archaeological treasures once meant for Romans who traded their familiar commodities between Mediterranean cities: Up came chalices, oil lamps, a lacrymatory—a miniature vial designed to collect human tears—and seven thousand amphorae, one of which was sealed and still contained wine (which, when we tasted it, was undrinkable).

On we labored into autumn, when mistral storms struck me as only an irritating distraction from our obsession. One day, the gales blew so savagely that they ripped our heavy anchor chain in two, sweeping the mooring buoy hundreds of yards away and leaving our anchor unmarked, somewhere 220 feet below the surface of a furious sea. Now we would have to interrupt our progress for the irksome task of reestablishing our moorings, beginning with the task of finding the anchor.

That very afternoon, two strapping young men, Navy combat divers back from amphibious action in the French Vietnam war, came out to the ship to introduce themselves to me. Jean-Pierre Servanti and Raymond Kientzy told me they so wanted to work on our financially straitened project that they would do so as volunteers. The two demonstrated their experience by diving at the site where we'd found the mooring buoy. Servanti even proposed an answer for finding our anchor: In the gale, the buoy had dragged its heavy broken chain along the bottom sands. We could simply follow the trail back to our anchor. The boy also enthusiastically proposed that we could beat the time restrictions necessary for such a search in deep waters—where humans must strictly limit themselves to a stay of no more than ten minutes—by descending not as a group but individually, in relays. In the howling wind, with the mess we were in, I felt moved by his initiative, and Servanti, eager to impress me even more, enthusiastically asked to dive first.

The next day, a few of us took a launch back to the site where we'd found the mooring buoy. Servanti plunged into the water and disappeared.

He never came back.

In the treacherous depths where the anchor lay, his heart stopped.

My conversations with this young man could probably be measured in minutes; I had shaken his hand, said, "Welcome to our family," and he was gone. I had, in effect, paid for a pile of old jars with a human life: It had been *my* decision to accept the manic demands of excavation, *my* decision to admire Servanti in his eager demonstration of his abilities, *my* decision to let him dive alone. My fault. Now I had to find his mother to announce her son was dead.

We made the grim final arrangements. We took the body to Hyères, Servanti's hometown. We buried the boy. And I returned to my ship, put my head in my hands, and swore this business was not for me. I decided to put an end to *Calypso*'s expeditions.

As I made my anguished plans, I received a cable from a diver who had worked weekends on our project but who had left on vacation just before Servanti's death. No simple words of sympathy, I thought ruefully, could stem my tide of remorse. But the cable's closing line caught me short:

May I have the honor of replacing Servanti to further your work?

[Signed,]

Besson

My dangerous work—my single-minded quest for those damned old jars—had cost one man his life. Now a second man attached such high value to those jars he readily offered to take on the job that had killed the first. Were these really only "old jars" after all? Old jars for which men were willing to die? Or was Besson's undaunted offer a sign that the ruins below were more than pieces of clay; that they were pieces of antiquity, pieces of a vanished way of life, messages from the human past that could inform the human future? I cabled him to come ahead. I would continue. Some risks are worth taking—because some goals are worth achieving.

My whole past turns on the hinge of this accident; all is either before it or after it. When I first began diving, acting as an individual, I had a right to choose any risk I pleased. Alone, I had been an adventurer. But as a leader of a team, I realized after Servanti and Besson, I would have no margin for negligence. I would have to change. I would have to become a *reasonable* adventurer. Besson had taught me that there are sound justifications for risking life; Servanti, that there is *no* justification for squandering life.

Together, the two men influenced the risk-taking rules I determined henceforth to follow: On our expeditions, we would not pursue risk; we would avoid it. Danger would be a mistake. Before a hazardous undertaking, we would spend hours attempting to anticipate all the deadly possibilities; then we would systematically try to eliminate them. Inevitably a risk would nonetheless remain: There is always an unknown, the essence of exploration. That one gamble—that literally calculated risk—would be the only risk I would allow my men to take. I would allow them to take that calculated risk, furthermore, only to achieve a worthwhile goal. Anyone merely addicted to thrills, anyone hungry to face danger for danger's sake alone, would have no place on *Calypso*.

My strict adherence to these principles has not sprung from any sense of "responsibility" I have for my team. To me, a leader motivated by his "responsibilities" acts less out of concern for his comrades than out of concern for his own interests, out of anxiety about the repercussions for his career should a serious accident occur under his jurisdiction. *Responsibility* implies liability. Yet how could a leader be "liable" for the life of a human being? What money could he pay? Instead of having the *responsibility* of safeguarding my men, I have the *privilege* of safeguarding them. I try to avoid reckless risk for the same reason I accept calculated risk. Life—improving the quality of all life, safeguarding a single life—claims supreme priority.

What "worthwhile goals," then, *do* justify placing oneself in jeopardy? What gains to life warrant risking death? The answers aren't so pat they can be found or written in some reference book. I identified the first goal for which I myself would accept risk when I was only twenty-six. I chose to take a risk if it could win me the opportunity to live, not merely to survive.

At the time, I was pursuing a lifelong dream: training to become a pilot for the French Navy. One evening on leave, as I sped along twisting rural roads to meet my fiancée in the village of Champagnat, my car flipped. As it somersaulted, the automobile smashed my body against the ground, then hurled me twenty-five feet into a field. I regained half-consciousness pinned to the earth, skewered at the left arm on the slender trunk of a sprouting tree. I tore the branch off at the ground and dragged myself to a nearby house. The farmer there, startled by my cries for help, found me lying in front of his gate. He called a doctor, who, in turn, rushed me to a hospital in the nearby town of Bourganeuf.

I had broken twelve bones. Four splintered ribs had punctured my lungs. Both arms were mangled. The right hung uselessly off my shoulder; I'd crushed its radial nerve. Doctors couldn't say if the nerve would heal or if I'd ever be able to use my right arm again. The tree had shattered my left arm, leaving me with four elbows, not to mention jagged bone ends poking through an open wound. Surgeons operated immediately, drilling in pins to hold the fractured bones together.

Two days later, the pinned-together left arm showed alarming signs of infection, a legacy of the dirty branch. My doctor wanted to act right away; if gangrene developed, it would kill me too swiftly for him to intervene. He presented me with what he regarded as a foregone risk-benefit decision: Amputate the arm, or die.

I saw only two risks, no benefits. To me, amputating the arm didn't seem like much of a vote for life. My dream of becoming a pilot had already begun to slip away when I had heard the uncertain prognosis for my immobile right arm. Amputating the left would extinguish my chance altogether. Life with one arm paralyzed and the other arm gone—life with no hope of achieving my goal—was not a life I could accept submissively. The fear of death did not nag at me; the fear of letting my chances slip away did.

I saw a third option. To the possibilities the doctor envisioned, that I amputate the arm or develop gangrene and die, I added the slim chance that I might *not* develop gangrene—and live. Waiting to see if the infection would heal posed the greatest risk, but it also offered the greatest benefit, my only chance to fight for the future I wanted. I couldn't forfeit. I refused the amputation.

For the next three days, my heart pounded in apprehension each time the doctor checked my infection. For the next ten years I exercised my arms. I regained full use of both.

I was lucky. But I provided my own opportunity for luck. I didn't want fear to prevent me from taking a risk that could give me a more meaningful life.

If improving the quality of life merits risk, then actually saving someone's life should mandate risk. Yet the logic that is clear on paper is not always so evident in reality, as I've learned from both my own instinctive reactions to danger and the unwritten rules of risk in foreign cultures. When I was in China in 1933, as a twenty-three-year-old midshipman aboard the French cruiser *Primauguet*, my duties included piloting a launch between the ship and Shanghai Harbor. One day as I pulled alongside the quay, hundreds of manual laborers stood passively at the dock's edge, watching with only mild interest as a man drowned. When I jumped into the water and dragged the poor fellow to safety, they began to laugh uproariously—at me! Only later did I understand why my actions seemed so hilarious: A widespread tradition holds that he who saves a man becomes father to the victim, expected to cater to his "son's" welfare ever after. According to this code, the human loss if a man dies is nothing compared with the nuisance if he lives.

Not until thirteen years later did I have the harrowing opportunity to see what I would do if I myself were dying alongside another man—and if persisting in efforts to save his life meant accepting the possibility of perishing with him. I faced that ultimate and awful risk-benefit choice during a dive into the flooded cave of Vaucluse, a water-filled shaft that burrows under six hundred feet of limestone rock in the cliffs near Avignon, France.

A few colleagues and I had decided to find the natural siphon, hidden somewhere deep within the cliff, that deluges the cave with water—an expedition I have already described at length in *The Silent World*. Briefly, I planned to dive into the cave with Frédéric Dumas—Didi—one of my closest friends. I'd known him since his days as a renowned goggle diver; we'd tested the early Aqualung together. Didi and I would descend into the cave like mountain climbers, tied together by a thirty-foot cordon of rope. A team member at the chasm's mouth had dropped a guideline with a pig-iron ballast, as far down into the cave's waters as the weight would freely fall. Didi and I would find our way into the abyss by following this guideline to the pig iron; then we would shove the weight off the ledge on which it had settled and into deeper recesses so that we could venture farther down. We described two signals to the men who would hold the end of the guideline at the cave's opening. Three tugs on the rope meant "feed more line," and six, "haul us out." Weighted with extra flashlights, an extra air tank, even an Alpiner's pickax, we had meticulously prepared for every eventuality—except the one that could kill us. Unbeknownst to us, a damaged air compressor had been used to fill our Aqualungs. Out of our mouthpieces hissed a lethal stream containing carbon monoxide.

The poisoned air had muddled my thoughts by the time we reached the pig iron. Didi, even more disoriented than I, bounced clumsily against the cave walls behind me. I found myself acting like an automaton, shoving the weight deeper as we had agreed. We forged farther down, our exertion compelling us to draw in long, deep swallows of gas.

Not until we had descended past a depth of 200 feet—having proceeded 400 feet into the slanting tunnel, away from the cave entrance and through the twisting bowels of the cliff—did I thickly comprehend that something wasn't right. The sickening anxiety I felt was not natural. Still I continued dully onward, compelled by the delusion that we had to find the secret siphon before we could leave. Didi, fumbling with his wet suit, which was filling with water, was himself nearly incoherent. I distractedly signaled that he should hold our guideline to the surface while I circled the cave walls looking for the siphon shaft.

As soon as I turned away, Didi let go of the rope. Our single line to life disappeared in the dark. Simultaneously, Didi's jaw went slack; his mouthpiece fell out. As he fumbled to replace his air hose, I could see his eyes roll back, their protruding whites glowing eerily in my flashlight beam.

I fought the fog settling over my own brain. For a flash I envisioned the nightmare awaiting us, in this watery tomb, if our air ran out. Desperately, I swung my light to scan for the guideline. Precious seconds passed.

I found it. Now how could we escape? I tried to shinny up the rope, with Didi dragging behind me on the cordon that tied us together. Our man above interpreted my first three grasps as a signal for more rope and fed down hundreds more feet of line.

By now Didi was almost a deadweight. I was nearing exhaustion. I tried again to scale the walls of the cavern, clawing for a hold; my fingers gave way, and weighted with Didi, I tumbled downward.

I dully realized I was dying. I could never climb out, not tied to Didi. The cordon and his body were ball and chain, fettering me in this flooded hell when I had only moments left. I could cut myself loose. I reached for my dagger. My fingers closed around its handle. I grasped the cordon that fastened me to my friend.

I couldn't bring myself to sever it.

I heaved again on the guideline—no more a signal, now a supplication. Above, they understood. They pulled us out.

Accounts about people who have been caught in such situations are never clear-cut. Those entrapped and dying alongside others must always feel torn between two equally powerful—equally understandable—but conflicting instincts: Save your friend; don't leave too soon. Save your life; don't stay too long. It would be gratifying, for a book, to be able to write about some complicated philosophy that influenced me while we were entombed down in Vaucluse. But my drugged brain was hardly capable, at the time, of any grandiose logic. What made me choose to risk staying with Didi? Just repugnance for death.

Of all the reasons that may justify risk, one of the most intriguing is also one of the most controversial: to fight for a conviction. During the Second World War, I had ample opportunity to see common citizens taking uncommon risks to defend their ideals. In the dark days of Nazi and Fascist occupation, I worked under a pseudonym in Navy intelligence with the Resistance movement in Marseilles. Mussolini's men gave us steady incentive for espionage; they didn't inflict much damage militarily, but they did economically—systematically looting our stores of strategic material, buttressing their costly war effort by stripping France of its future.

One day my assistant intercepted some military telegrams that listed a series of numbers. We both guessed that several Italian "Armistice Commissions," scattered throughout the occupied south of France, were exchanging encoded information detailing plans for coming raids on French resources.

Given that the telegram we had in hand was destined for the Commission of Sète, I guessed the office there probably had secured some sort of decoding documents in a safe. I knew my superiors would never allow me, as a French officer, to burglarize the commission myself. But I felt certain that I would easily find French nationalists in Marseilles who would take whatever chances were necessary to help us lay hands on the safe's contents.

I was right. The French cleaning lady who swept out the commission premises risked imprisonment for complicity by sketching a blueprint of the office. She pinpointed the position of a large safe, manufactured by a well-known company, Fichet. A politician who also served as a lawyer for the Marseilles underworld introduced me to a gangster he had represented in court; the thug was not only a patriotic Frenchman but also spoke fluent Italian. He accepted the crucial role in our burglary: He would impersonate an Italian officer and penetrate Fascist headquarters with cohorts I would find for him.

Even the thug's mistress was willing to incriminate herself for collusion in our conspiracy. She seduced the real Italian captain in charge of Sète headquarters. As he slept, she picked his pocket and made a wax impression of the key to the safe. I recruited an expert locksmith from Fichet, the safe's manufacturer, to sculpt a key from our wax mold.

Opening the coffer, however, would require not only a key but also a combination. The Fichet locksmith agreed to join our heist so that he could personally try to coax the safe, with its combination dial, into opening. A quartermaster in the French Navy accepted the third role for the break-in, that of photographing the safe's contents. I found a black Citroën car for them, just like the one the Italian officer drove, and I forged copies of his license plate and official windshield symbol. I chose a date and scheduled the burglary to begin at midnight.

The day of the raid finally arrived. In the morning, the gangster came to see me. He couldn't go through with it. He was torn between a fierce dedication to our cause and his almost nauseating fear. "Finally, I can't do it," he said. "My heart would fail." He actually swore he would even kill someone in the open street if asked—he would do anything, anything but the burglary.

After all these weeks of preparation! Now, at the last minute! There was nothing to do but go myself.

The hours that day spun by too quickly. Gambling on the chance that the sentinel on watch would not ask for identification if our group looked familiar, I decided to mimic the style of the Italian captain who headed the commission; I'd wear a light gray suit and a fedora, as he did most of the time. I disguised my quartermaster in the flashy costume of a Guardia di Finanzia, with its renowned *piuma sul capello*. I hoped the Fichet specialist, short and thin, would pass as unremarkable.

We each needed only one last accessory: a weapon. I forbade guns. If the Fascists captured us and saw we had no firearms, they might send us to a concentration camp instead of executing us on the spot. I handed out truncheons, strips of rubber studded with lead, with which we could stun any intruder and try for escape.

Just before midnight, in total silence, the three of us drove the Citroën to Sète and pulled up directly in front of the headquarters. If anything was going to get us through the night, it would be ostentation and insolence, not stealth. Faces frozen in nonchalance, hearts drumming, we strolled to the front door, where a real member of the Guardia di Finanzia stood as sentinel.

If we had to speak to him, we were through. None of us knew enough Italian. Without a word, I dug into my pocket and found the door key I'd borrowed from my friend the cleaning lady. I slipped the metal shaft into the lock and prayed that she'd given us the right key. The bolt yielded. I opened the door. The sentinel, in recognition of officers, presented arms.

We sauntered into the office and slammed the door purposefully behind us. Our locksmith quietly hurried to the safe and tried the key we'd copied from the one in the Italian officer's pocket. It worked. Then the locksmith put his ear against the combination lock. For forty-five minutes—oh, what a forty-five minutes!—we listened to the staccato clicks of that spinning dial as the locksmith feverishly persisted in his forays.

At last the heavy metal door swung open. There was the code book. My quartermaster, in his beplumed Italian disguise, grabbed it and instantly began photographing, page by page. I looked deeper into the safe. A stack of envelopes lay inside, one ripped open and the others sealed: cipher keys! The Italians were not just encoding messages; they were enciphering them, scrambling them a second time according to keys they changed each month. If we were to confound future Fascist raids, we had to photograph the key for the current month, in the open envelope, as well as the keys for future months in the sealed envelopes, without leaving a trace.

We had no way of knowing, of course, that our gangster acquaintance and his girlfriend were meanwhile having some trouble with the roles they had offered to play for the evening. They were diverting the real Italian officer in charge of headquarters by taking him on a round of local nightclubs. He announced he had grown bored. What, our two accomplices asked, would he like to do next?

He wanted to stop by his office.

Why not, the woman suggested seductively, take her home with him instead?

He wasn't interested.

The officer walked out of the club, his two companions following numbly behind. Together they went to the Italian's familiar black Citroën and piled into the car. The officer reached into his pocket for his keys, started the engine, pulled out of the parking lot, then onto the road. My gangster acquaintance groaned. Abruptly, he clutched his side. Then he doubled over. His voice choking in pain, he gasped that he must have ruptured his appendix. "Take me," he mumbled, "to a hospital."

His shrewd theatrics bought us time. At exactly four A.M., after four hours inside the guarded office, we faced the final challenge—leaving. We opened the door, stepped out authoritatively, and slammed the door behind us. A sentinel presented arms.

We delivered our film of the code and cipher keys to the French Naval Central Intelligence. We were never told how—or even if—they were used.

Such escapades make it easy to overlook a key point in the subject of risk taking for convictions: Whether a risk is reasonable depends on whether the convictions are reasonable. One nation's loyalist is another's kamikaze; one religion's martyr is another's terrorist. That the leveling perspective of time can sour an ideal is evident in the case of those medieval fanatics who believed the quickest way to heaven was to kill someone who adhered to a different religion. Today we still evoke the name of their sect, but not as a metaphor for fervent faith. They were the Assassins.

Yet while a mass readiness to die gives a tyrant his most terrifying weapon, a people's will to stake their lives on their beliefs also forges the most formidable weapon *against* tyranny. The great religions have endured not because zealots spilled the blood of anyone who resisted their creed, but because the faithful were willing to spill their own blood, to aid others, in accordance with their creed.

The risk takers whom I most admire—those whose achievements have ennobled the world, whose accomplishments endure—have been the kamikazes of peace. They face danger not to wreck havoc on their enemies but to contribute to the welfare of their fellow human beings. If personal gratification from headlines were all Jean Mermoz had been after, he could have retired from risk taking in 1930 as the idol of French youth during my childhood for being the first to fly directly from France to South America. Instead, Mermoz turned his one-time feat into a weekly act of courage for the next six years, finally losing his life over the sea. Why? To deliver the mail, to put people on distant continents in touch with their families.

Roald Amundsen had no need for additional personal glory when he risked his life to save General Umberto Nobile, the Italian aviator lost somewhere on the Arctic's expanses. Amundsen had already won world renown as the first to reach the South Pole, then again as the first to lead an airship expedition across the Arctic continent, accompanied by Nobile. Nobile, who had designed their dirigible, bitterly envied Amundsen's resulting honors and in 1928 undertook to lead a poorly prepared mission in his own Arctic flight. When the Italian crashed, it was Amundsen, nearly bankrupt at the time, who sold his belongings to finance a rescue of the man he felt he had driven to the ice. A Russian pilot eventually rescued Nobile; a Soviet icebreaker later saved eight of the sixteen men Nobile had left at sea. No one ever saw Amundsen again: A lone fisherman watched the doomed plane disappear in the fog as the famed explorer set out to save the life of the man who loathed him.

So too did the physicians James Carroll and Jesse Lazear have nothing other than altruism in mind when they followed bacteriologist Walter Reed to Havana's prison wards in 1900, during a yellow fever epidemic in Cuba. Because no one knew how the fever was transmitted, Carroll deliberately dared to submit himself to mosquito bites and soon fell severely ill. Lazear also willingly accepted risk, attending fever victims in the mosquito-infested wards, where he too was bitten. He died. The sacrifices of these two men confirmed that the insects spread yellow fever; authorities sprayed pesticides, and in just ninety days the epidemic ended.

Is this heroism? Those who accept such risk probably would call it decency. I doubt that even Mermoz, Amundsen, or Lazear would have attached much significance to the fact that they would one day die for their convictions. They believed it was important to *live* for them.

A worthwhile goal alone, however, does not in itself justify risk; the means of executing a dangerous mission must also be sound.

While I would not knowingly take a chance with the odds stacked against me, I have unfortunately taken chances unwittingly, stumbling on the first category of unacceptable risk: what I call risk by ignorance—the dangers that a prudent adventurer would never choose intentionally but that he cannot avoid merely because he treads unknown territory. Risk by ignorance is exploration's occupational hazard.

One such mistake my men and I made, with no idea we were making a mistake, involved a dive on Clipperton Island, an uninhabited atoll off the Pacific coast of Mexico. An inactive sunken volcano rises up from the floor of Clipperton's lagoon, its crater gaping open under the pool's surface. We used our echo sounder to plumb the depths of the crater's mouth; sixty feet down, our data indicated, the opening was sealed off by some kind of false bottom, which floated eerily above an even deeper chasm.

One of my cinematographers made a brief dive into the crater; he reported that the water grew inexplicably warmer as he descended. Thoroughly puzzled, I organized a diving expedition, taking the precaution of asking everyone to dress in our polar watertight suits—which would cover faces, hands, ankles—to protect not against cold but against heat.

Sure enough, as we submerged and descended, we felt distinctly warmer water. At sixty feet, we solved the mystery of our echo sounder's shadowy false bottom. Decomposed plant matter, which for years had fallen from the surface and putrefied, had formed a blanket of debris all the way across the width of the crater. As we approached, we stirred the dormant waters, and the rotten sheet billowed like a massive blister. One by one each of us dove through the decaying layer. Below we found a lifeless Hades where no visible creature survived. Caustic vapors penetrated our masks and stung our eyes. Our gloves leaked, and the liquid scalded our hands. At once I understood: We were awash in hydrogen sulfide, the acid produced when rot consumes all oxygen in water. The immutable laws of diving required us to wait, blinded and burning, at decompression stops on our ascent. We emerged with hands scorched and eyes weeping from exposure to an acid so concentrated it had turned the bright-yellow paint on our air tanks to olive green.

If an absence of knowledge is dangerous in risk, a little knowledge is outright deadly. When I came to know a modicum—but far from enough—about sharks, I blundered into unreasonable risk's lethal second category: risk by overconfidence.

The first time I faced a shark, in 1939 near the island of Djerba, off Tunisia, the numbing shock of seeing the beast right before me prevented me from doing anything foolish. As we continued our voyages through shark-infested tropical seas, however, we grew accustomed to diving among the carnivores, to admiring the sleek beauty so many have found hideous. Soon we discovered we had but to swim purposefully toward the hulking animals and they would turn tail and flee. Only the reputation of the great white shark, which specialists call the most malevolent of man-eaters, continued to disturb us—until our first encounter. When a twenty-foot creature, the beast of our nightmares, once rocketed before us, he did not notice us at first. But as soon as he registered our presence, he voided his intestines and bolted, leaving us laughing nervously in a cloud of excrement.

The emotional wash of such effortless victories made us arrogant, and before long, our arrogance blossomed into gleeful contempt. These reputed brutes—so strong they can propel their expansive forms through miles of water without appearing to move a muscle—we were mightier than they! Inexperience inspired recklessness. When we wanted to film "lively" sharks, we pulled their tails. We made derisive jokes about the cowering impostors. In more timid days, we had designed an antishark cage; now we disdained it, scornfully referring to it as a "human zoo."

It had not taken us many years to become overconfident divers. One day on board the French Navy's diving ship *Élie Monnier*, as we sailed from the first bathyscaphe expedition in Dakar to the Cape Verde Islands, we encountered a huge school of bottlenose whales. The weather was fine, the open ocean dead calm, the whales exceptionally placid. Didi and I decided to dive and film the whales sounding vertically, disappearing in the deep blue to catch their invisible bounty. By the time we had equipped ourselves, the whale pack had thinned as the behemoths had started dispersing. The crew lowered a ballasted rope overboard, as a safeguard, and Didi and I jumped into the water.

Dismissing the lifeline, we swam swiftly away, hoping to approach one of the laggards. We filmed its graceful plunge; then it vanished, leaving us behind, two men wafting in crystal clear water. It was not long before we realized we were not alone.

A shape was materializing out of the darkness of the depths—the shape of a shark we had never seen before. We later came to call him "Lord of the Long Arms," after his Latin name, *longimanus*. Unlike the torpedo-sleek gray sharks that had burst upon us before, this creature—with his long flippers extended like the wings of a fighter plane—soared. We couldn't resist temptation. We didn't bother to look for the safety guideline or wave for the men on the ship as a precaution; we simply sallied forth after the newcomer—just off to scare another coward into flight.

We pursued "Lord Longimanus" downward. At last he swerved into a wide circle. I filmed as Didi clowned, swimming side by side with the shark, even reaching out to tweak him.

But this beast made no move to retreat. He circled. He fixed us both in an expressionless stare.

Sickeningly we realized that his unremittant circling was not just aimless behavior. He had targeted us as his prey in a dance of death.

We managed to surface briefly to discover that our ship, following the whales, was bearing off from us and had lost sight of our bubbles. Didi and I were hundreds of miles from any land, drifting away from our ship, suspended in a three-dimensional universe, open to attack from every direction.

Abruptly the *longimanus* lunged for me; I desperately thrust my camera forward, smashing his muzzle. He was unfazed. He returned to his inexorable circling. Now two fifteen-foot blue sharks shot into the circle from out of nowhere. With their bands of nerves from snout to tail, they could sense even a shiver of fear undulating through the waters. The blue sharks joined the *longimanus*'s unrelenting carousel, as Didi and I twirled dizzyingly at its axis, in what seemed to be an increasingly futile attempt to keep them all in view.

We made a break for the surface, popping above to flail frantically for the ship; but by now we could see the *Élie Monnier* only faintly, some three hundred yards away. From below a shark thrust upward, aiming to pluck off legs that dangled from the surface like bananas. We fended him off. Frantically we formed a human pinwheel, each man's face positioned so that he could watch the other man's feet, as we somersaulted to the surface to signal alternately for help. The fiends below circled, indefatigably, awaiting our final and fatal exhaustion.

Suddenly we were saved. A dinghy arrived. Friends snatched us out of the water and returned us to the ship.

The more experience I gained with sharks in subsequent years, the more they shook my one-time overconfidence with them. Twice, from our antishark cage, I have witnessed their feeding carousel. I have also observed *longimani* as they trailed packs of whales in programmed pursuit, primed to devour the weak or the sick. I've watched as a lunging shark took one bite out of a whale and scooped out a bucket-sized piece of the behemoth. Remembering my hubris in my own too-close encounter with a *longimanus*, my blood has run cold.

My blood also ran cold when the life of my friend Luis Marden was almost lost to the razor-sharp teeth of agitated sharks—but to this day I'm not certain if it was Luis or the sharks that chilled me most. We were filming *Silent World* in the Red Sea; *Calypso* was anchored near the Farasan Islands. The National Geographic Society had sent Luis, its renowned photographer, to join us. One day, the sea suddenly started boiling around our ship. Dozens and dozens of large sharks, in a terrifying frenzy, were biting at anything afloat, even sticking their gaping jaws into our wooden hull, even cutting one of their own kind in two. While we gaped, sickened, at the fury, Luis hurriedly donned his Aqualung, prepared his camera, and headed for the diving ladder. I jumped to the fantail, barring his passage. Luis shouted at me. Shaking with horror, I refused to argue and called for help to neutralize him. He was shrieking: "The picture of my life! The picture of my life!"

The picture of his life would have been the last picture of his life.

Such risks by overconfidence may, at least, provide those who commit them the opportunity to learn from their mistakes—if they survive their mistakes. Unfortunately, overconfidence is not the last of the factors that characterize senseless risk. In a life spent at sea I've met men driven to take risk for money; risk to show off; risk for renown. With each haphazard success, the allure of risk becomes more magnetic. Too often the risk taker finally takes risk for nothing more than the sake of risk itself—like the notorious rogue of Corsica, the coral diver Toussaint Recco.

As we began work on a documentary about the Mediterranean's red coral, Corsicans told us tales of Recco, their most brazen risk taker of all. His Christian name, Toussaint, literally means "all saints," a congregation everyone suspected Recco would meet early, given the insane chances he took.

Divers had already hammered and crowbarred red coral, a precious species coveted by jewelers, out of all shallow waters; spurred by the increasing price of the vanishing booty, they were pressing on deeper and deeper. Recco won his reputation—as well as a king's ransom—by fetching branches that remained in the deadliest depths.

I've seen that most drowning accidents do not befall beginning

Aqualungers but claim the lives of those with some experience, those who think they have mastered diving. Thus I wasn't surprised to find that Recco, an accomplished diver, first acquired his taste for risk through overconfidence. He believed he had outwitted natural laws, boasting that he had even overcome nitrogen narcosis—the inebriation caused by breathing nitrogen, a major constituent of air, under the pressure of deep waters. During our first deep dives with the Aqualung, we had experienced this euphoria, akin to the effects of chloroform, and had come to respect its treachery. We dubbed it "Rapture of the Deep." Rapture is not an instantaneous killer but rather an insidious one; it lures a diver to deeper waters even as it robs him of reason. At 40 feet down, I personally feel a giddy merriness; at 200 feet, a delusive heedlessness. In experimental dives below 200 feet, nitrogen has so addled my brain that I have lost my sense of up and down, felt vaguely close to losing my instinct for self-preservation. One of our colleagues in early Navy test dives in fact lost his life. Heading back for the surface ends the inebriation. Adding the appropriate dose of helium to the mix of air will prevent the intoxication altogether, but this so-called mixed-gas diving itself poses challenges. Mixed gas is expensive and not available everywhere; for us, the unwieldy canisters are difficult to transport to far-off destinations on a little ship. When we've had to forgo mixed gas on deep dives that flirt with the Rapture regions, especially after the death of Servanti, we've taken special precautions and followed decompression rules to the letter.

Yet Recco—who had probably never heard of helium and who disdained what he most surely considered our crazy complications—not only plunged to 300 feet on air alone but also stayed in this perilous zone as long as it took to harvest all the branches he wanted. He scoffed at the rules of decompression as well, improvising his own diving code as he surfaced. At the brief decompression stops to which he grudgingly conceded, he arranged to meet a curvaceous beauty-parlor stylist who liked to dive on her vacations; they passed time by sharing his mouthpiece. At first I was dumbfounded as to how Recco had escaped the crippling effects of the bends, but then I saw that in fact he hadn't. Back on solid ground, he limped. To give himself even more time in hazardous depths, he overpressurized his air tanks, turning them into miniature bombs. With great braggadocio he displayed the remains of one tank he had accidentally exploded; its metal sheathing had burst like an overblown balloon, showering knife-edged shards of steel around him.

What could so propel a man toward disaster? During the filming of our documentary, an avaricious coral broker who had made a fortune buying and reselling Recco's finds declared that he, for one, felt great affinity for what he perceived as the diver's motivation: money. The more risks Recco took, the richer he became. Yet I watched Recco's cravings burgeon—first from coveting money, then to coveting coral. He spent hours perusing the ton-and-a-half collection of branches he had hoarded in a shed. Recco so savored the depths that I think he even craved the drugged pleasures of Rapture. I believe he actually became addicted to it and then finally addicted to risk itself. The cheap thrill of staking his life—risk for risk alone—became his opiate.

Years after we met Recco, his friends gossiped over what they called his ironic end. He died a violent death, but not through diving. His brother-in-law shot him dead in an argument over a garden hose. To me, Recco's death by murder offers no irony. I found him charming—his risks even mesmerizing—but still I feel he died as he had lived: for nothing.

To me, the public's fascination with such recklessness seems misplaced. When a braggart boasts in headlines that "danger is my business," it seems to me he might as well add, "and my specialty is stupidity." A daredevil displays neither talent nor cunning when, simply to see if he can narrowly avert disaster, he fumbles through a hazardous undertaking. The more demanding goal is to dare without danger, to achieve a worthwhile goal by systematically minimizing, and thereby mastering, risk.

What are the methods, then, of reasonable risk taking?

Early on I discovered the foremost imperative for confronting risk in a calculated way: Suppress fear. Risk itself is only a threat. Fear is an enemy. In the face of a danger that requires clear-witted reaction, fear clouds judgment.

Even though fear is not a noble sentiment, it is a normal one. I have never been able to rid myself of the emotion, but I have learned that I can at least control fear if I can gain any control over the situation itself.

I was paralyzed by fear, for example, when the Americans bombarded the Gare de La Chapelle in Paris in their effort to cripple occupied France's transportation of strategic materials. I had no way of protecting myself, my wife, or my sons. The bombs fell everywhere and anywhere: Why not next on us? I could only huddle at home with my family, sick with terror. Yet my memory of swooping Italian planes trying for a direct hit on my ship the *Dupleix*, harbored with the French fleet in Toulon, includes no recollection of fear. Again the bombs screamed down and exploded all around me. But I manned the antiaircraft guns and thus exercised at least some control over my fate. The heat of action left me no time to be afraid.

The menace of fear extends beyond its power to paralyze. Fear also fires the aggressiveness of an assailant. Animals smell fear; they react to its presence as an invitation to instant attack. The appearance of calm often magically deflects a predator. I wouldn't rely on serenity as my sole defense against a carnivore, but I learned, during the 1930s in Cambodia, that it works at least as a last resort. A fellow Navy officer and I were sightseeing in the jungle, picking our way through the overgrowth that some five hundred years before had swallowed the lost city of Angkor, crowning glory of the Khmer dynasty. We came upon the remains of a small temple, its stones still etched with symbols that Southeast Asian kings had hoped would magically turn them to gods. My friend lingered in the little temple, but I dreamily wandered on, immersed in the beauty of a jungle that had ultimately overcome the once glorious civilization.

I never expected the black panther that stalked onto my path. Panthers in this jungle had been known to kill dogs and livestock—as well as humans—during attacks so agile that leftover flesh and bones had been found among the vines up in the forest canopy. I slid my back against a tree, to barricade myself at least from behind. Then I fixed my eyes directly on the panther's. Her muscles rippled as she looked me over. I held my gaze. She padded majestically off. Of course I had been afraid. But I am convinced that had she sensed it, I would not be here more than half a century later telling the tale.

The most common risks my team and I face, in fact, involve encounters with just such unpredictable animals. In managing those risks, fearlessness helps keep our minds clear, but relying on fearlessness alone would amount to foolhardiness. As we can't reduce the frequency of nature's surprises, we increase the thoroughness of our preparations.

Our past experiences have inspired our present array of precautions. The black panther incident gave me one idea for protecting ourselves from sharks. When a diver hangs suspended in an open sea, sharks can approach from every angle. Placing my back against a tree in Cambodia reduced my field of surveillance for a panther attack; it prompted our policy of positioning two divers back to back for better scanning of shark-infested waters.

The *longimanus* episode itself inspired another useful precaution. When in desperation I rammed the lunging shark with my camera, the blow sent *me* tumbling backward. For at least a moment I had pushed myself out of reach. After that incident we fashioned the "shark billy," a simple wooden staff with nail tips studding its far end. The nails don't hurt a shark but keep the stick from slipping off its hide as we shove ourselves back. With billies, two divers back to back can face two sharks.

But any number of divers with any number of billies can find themselves no match for more than two sharks. In shark-infested waters, I levy an absolute rule: Our divers must lower antishark cages, those aluminum- or steel-barred boxes Didi and I once ridiculed as human zoos. Divers need not necessarily use them; but if the sharks' mood unexpectedly shifts from wariness to fury, the men can retreat into cages and slam the doors. We can winch the divers up in safety. As an extra precaution, I often install an undersea hydrophone and a TV camera in one of the cages, as well as closed-circuit receivers in our chart room, on which I can monitor the confrontations below and respond immediately to any developing threat.

Such preparations made it possible for me to watch my son Philippe relive my long-ago encounter with Lord Longimanus. Philippe registered all the vivid first-person emotions I had experienced but displayed none of the recklessness. We had decided to study shark migration around the reef of Dahl Ghab Island, off Sudan, by "labeling" the beasts with numbered tags, which we would plant into their fins like banderillas with a thrust of a long lance.

We lowered two cages. In the first, a diver held his tagging lance and a sack of fish morsels to lure sharks. In the second, where I'd installed the hydrophone, Philippe carried a camera to film our corrida in the sea.

From below, Philippe loudly announced over the hydrophone: "Arrival of a *longimanus*." We were not to hear his voice again during the dive. We watched on the closed-circuit receivers: Here came the monarch, large, majestic, insolent, surrounded by a court of six pilot fish. Lord Longimanus circled the mob, sure of himself, indifferent, not about to mix with the vulgar commoners. Suddenly, in a lightning flash: A middle-sized white-tip shark seized a large fish head that the divers offered; the carnivore was at once attacked by the bigger sharks.

In a movement so fast we could barely follow it, the *longimanus* charged, dispersed the lot, and swallowed what was left of the bait. In a single stroke he then whipped around, swooped on the hydrophone that hung from *Calypso*'s hull, snapped it up in his mouth, spit it out, and attacked the TV camera, gnashing it in his teeth. He savagely lunged for Philippe's small aluminum cage, seizing the bars in his jaws and shaking the whole apparatus ferociously, with Philippe rattling inside like a human marble in a tin can.

Philippe then watched in horror as the shark tore at the rope on which his cage hung from the ship. Watching the scene from the television above, I ordered the crew to winch both cages back up to the ship. When my trembling son managed to extricate himself from the cage, he stumbled out on deck only to see me laughing helplessly. The *longimanus* had terrorized him as much as its ancestor had once terrorized me. But tragedy had turned to comedy because, with our precautions, Philippe's melodramatic expression had been only a reflex, not a product of actual unreasonable risk.

Even precautions, of course, can only partially protect against the unforeseeable, as we discovered very late one night when Bébert Falco, my closest collaborator on *Calypso*, and I took the diving saucer, equipped with an ultrasonic telephone, along the sheer underwater cliffs of the reef of Shab Arab, in the Gulf of Aden. As we submerged, dozens of tiny green lights perforated the black of the water—shark eyes glittering in our headlights. In every direction the sinister glinting lights proliferated. Soon some six dozen sharks were banking around us, below us, above us. One after another they swept by before us, their dance charged with tension, in the dead of the night. Our narrow portholes prevented us from filming the full saraband, so we phoned the ship and asked some cameramen to dive, cautioning them to lower a cage.

By the time two divers were ready, the sharks had gone wild, maniacally lunging at anything and everything. The men entered a cage on board, before it was even winched into the pandemonium below. Falco and I watched from the saucer as the cage, its two men securely inside, was lowered into the demented pack. We were absolutely baffled by what we saw next. The two divers safe inside the cage had dropped their cameras and had begun to whirl, to wave their arms, to slap frantically at their feet. I stared open mouthed at the two dervishes. The cage went up; the divers evidently had sounded its alarm. Bewildered, Falco and I ascended in the saucer and docked it aboard *Calypso*, only to climb out onto a deck spattered with blood. We raced to the mess, where the two divers lay in agony on a table, their ankles swathed in bandages soaked by widening, wet splotches of red. Safe from the sharks outside their cage, the men had been trapped inside with clouds of near-invisible sea gadflies. Thousands of the tiny clawed crustaceans, ravenous as piranhas, had mercilessly torn off strip after strip of the only flesh exposed on the divers, on their ankles between the bottom of their wet suits and their flippers. An unanticipated hazard had once again threatened our men. But a precaution—our alarm system—at least had delivered them from it.

Time after time such unexpected perils demonstrate that no matter how much a risk taker knows about nature, he never knows enough. Even animals we've come to know as peaceful have, when cornered, become unexpectedly violent. On the Pacific island of Guadeloupe, we camped for weeks in tranquility among the hundreds of elephant seals that come to this protected sanctuary to mate. We knew that the males fiercely battle each other for females by slashing out with their formidable fangs, using them like swords to cut their competitor's enormous snout to ribbons. One thrust of a tusk, backed up by an elephant seal's three-ton hulk, could kill a man instantly. But it became evident, as we walked the beach, that the lumbering giants' sheer mass rendered them incapable of giving humans the chase. Even when they showed some sleepy anger, they "charged" us only by rearing upward and heaving forward, dragging their heft caterpillar-style until the effort exhausted them.

What we hadn't fully anticipated was the animals' supremacy in water, where their weight turns to might and their flabby torpor to agility. We thought we could film mating elephant seals if we surprised them in a shallow pool, so our divers stealthily swam toward a pair, in the process mistakenly blocking the narrows that served as the pool's only exit. The cornered bull veered in the water with astonishing ease and hurtled toward the divers in a rage, crushing a pressure-proof camera in a single bite. Then he and his mate were gone. Only a cloud of semen from the interrupted intercourse, its milky trail dissipating in the current, left any trace that they had even been there.

We thought we had mastered animal unpredictability by the time we started a documentary on hippos in Zambia's Luangwa River. As we set up camp nearby, the wallowing behemoths behaved exactly according to their reputations, lethargically lolling in the water, relieving the oppressive burden of their tonnage with their buoyancy. This time we paid greater heed to the risks of mating season, which turns a male hippo into seven thousand pounds of flesh and fury. His two bottom teeth and two top teeth, which fit together like the ridges of shearing scissors, can saw a man in two.

Sobered by our previous experiences, we built a remote-camera raft so that we could film the hippos from afar. We fastened one camera to the bottom of the craft, to film underwater, and another camera to the top, to record surface action. Then each of two men took one of the two leashes we'd attached to the raft and walked along opposite shores, guiding the craft as it floated out on the middle of the river between them. The camera rig worked flawlessly for minutes. The farther the men advanced, the more they crowded two hippos into a narrowing stream. One hippo reacted by plunging under the leashes to win freedom in the wider part of the river. The second hippo, still corralled, burned with rage; with no one to expend it on, he exploded into a stampede toward our raft and dived. Entangled in the leashes, he erupted up from the water and smashed down on the rig. He seized the shining underwater camera in his cavernous mouth and crushed right through the durable casing as easily as a trucker would crumple a beer can. I had to pay for his rampage by replacing yet another expensive camera. Better that than to pay with a life.

While "daring without danger" requires arduous preparation, the most exacting skill demanded in risk management involves deciding when not to dare at all. After investing time, effort, and aspirations, it can be tempting but fatal to become wedded to your plans. On one unforgettable occasion our team members abruptly aborted their mission; the threat they could not reasonably face was posed not by the unpredictability of a wild animal, but—for the only time in my career—by the unpredictability of humans themselves.

We had spent two years exploring the Amazon basin, following its waterway from mouth to source, filming its gold mines, its dams, the effect of encroaching civilization on the wilds. To complete the investigation, I decided to send my son Jean-Michel and four team members deep into the jungle. They would record sounds and sights of creatures already vanishing from one of the last virgin forests on Earth.

All of us had heard bloodcurdling tales about the jungle's Indian tribes. Only several months before we arrived, one of the tribes reputed to be most ruthless—the Yanomamo—had captured and slaughtered four miners prospecting on Indian hunting grounds. People living in Brazilian cities cited such tragedies to illustrate the natives' savagery; to me the stories testified instead to a kind of integrity. The Indians were defending more than their lives; they were defending their way of life.

As much out of respect for the Indians as for the safety of my men, I hired a well-known *sertanista*, an Indian-contact man. Once employed by the Brazilian government to communicate with tribes before the proposed construction of a jungle highway, our *sertanista* spoke several Indian dialects. When I asked him to choose a camp for our film team, I so trusted his expertise I gave him nearly free rein, specifying only two inflexible criteria: The campsite must be chosen far away from any Indian settlement, and under no circumstances were our men to come into contact with any Indians themselves.

Scouting the wilds with me by amphibious plane and helicopter, the guide proposed to pitch camp on a sandy islet in the Río Padauari, near the border between Brazil and Venezuela. I accepted his suggestion. The location would afford us a possibility we would not enjoy elsewhere: If our men were to have set up camp under the canopy of a dense uncharted jungle, in an emergency we would never have found them. But if they settled on a river, they could retreat along the course of the water; we could follow the stream and find them within two days.

The site posed its risks. It would place the men too far away from reliable radio contact. We needed two pilots to fly the men to the camp— one, to operate our amphibious plane to a stretch of water straight enough for landing; the other, to use our helicopter to shuttle two team members at a time the rest of the way to the small site. Each trip involved flights over forest too thick for us even to find the wreckage of an accident. We compensated by taking every precaution we could conceive. We equipped the aircraft with Air Force survival vests, first-aid gear, flares, machetes, and food for six days. We arranged for the pilot of the amphibious plane to circle above camp every other day to communicate with the men via walkie-talkies. Once settled at the site, the *sertanista* issued two other cautionary rules: No one was to venture into the jungle alone, and every group that entered the forest had to carry a machete and a gun.

The first fateful error seemed only an irritation at the time. The *sertanista* disgruntled Jean-Michel one day by shooting a tapir. Not only was my son disturbed by the unnecessary kill; he worried that the ringing shot would scare away the very wildlife the team had come to film. In the days that followed, the bodies of birds periodically fell heavily through the trees, as the guide ignored Jean-Michel's prohibitions and persisted in his hunt.

The *sertanista*'s rifle practice was forgotten as other apprehensions edged into the minds of the men. One afternoon, as they followed the guide through the forest, they had progressed not more than one hundred yards from camp when they noticed machete gashes in the trees— trail markings used by Indians as they wend through lush jungle. The guide grazed his fingers lightly over the darkened slashes, examined the new buds that already sprang from some of the healed scars, and pronounced the Indian trail some six weeks old.

Days passed. The men were once again delving into the jungle with the guide when they came upon an empty Indian camp, ravaged by weather and dilapidated by time. The guide insisted that Indians never return to use a camp a second time. The fact that the Indians had abandoned the crumbling site long ago was obvious, but not altogether reassuring. Jean-Michel noticed that it had been built in the characteristic fashion of the Yanomamo, the merciless tribesmen who had executed the four miners for trespassing on Indian hunting grounds. My son's mind spun with memories of the noisy helicopter landings, the fishing equipment they had brought—and worst of all, the gunshots, the animals that our guide had killed. Or poached? On Yanomamo territory? From that moment on Jean-Michel felt they were being watched.

On the ninth day of their planned two-week stay, the men once again set out into the undergrowth, ducking vines and listening for the subtle sounds of animals they wanted to film. They found more hack marks on the trees. The guide squinted at the splintered wood of the various gashes: one week old. Three days old. The men were silent. The green pulpy wood exposed by the next gash still glistened with a trickle of fresh sap. It had been cut the day before.

The only way to manage the risk that those gashes forewarned was to avoid it altogether. Jean-Michel decided to evacuate immediately. The men spent an unnerving night breaking camp, trying not to jump at every crack of any twig. Dominique Sumian, the strong but gentle giant of our team, has long been experienced in working in dangerous conditions. Jean-Michel assigned him the task of overseeing departure. Because the helicopter pilot had room for only two men at a time, Dominique made a list of the men, giving priority departure first to those with children and then to those without children but with wives. He divided the gear into two piles. The first contained the essential equipment they needed to take with them; the second, everything else—clothing, boots, leftover food—which Dominique placed at the tree line where the beach and jungle met as a show of generosity to those silently watching from the forest. One last problem: Because of the two-passenger limitation, while the helicopter made its second-to-last shuttle out of the camp and then its return trip, one man would have to wait at the site alone. Again Dominique volunteered. He had perfect qualifications. He knows how to handle a rifle. And he knows how to handle fear. As Dominique waited for the helicopter to return, he walked to the riverside, peeled off all his clothes—using them to hide his rifle and ammunition on the beach—took a swim, and even shaved. A naked man bathing in a river, he reasoned, had stripped away every sign that he posed a threat—and every sign that he registered any fear. Yet Dominique projected only the image of calm. As he floated on the warm water, he heard the sinister whistle of machetes slicing through the jungle just a hundred yards away. At last that sound was drowned out by the rotors of the returning helicopter. When all the men were finally assembled on the plane, its pilot took off and made one last circle over the campsite. Below, they saw Indians gathered on the beach, converging on the pile of goods left behind.

Even more harrowing are those risks that offer no escape. For me just one experience exemplified this extreme risk requiring the most meticulous of precautions: diving into one of the hundreds of chasms that burrow in unfathomed, interconnecting passages under the Caribbean in the Bahamas.

My entire team was convinced that a successful dive would yield inestimable rewards. Those who dared to enter these caverns could read the account of creation written in rock. Dr. Robert Benjamin, a chemical engineer by profession, an undersea cave explorer by passion, tantalized us with stories about one crevice so deep he had been unable to probe all its mysteries. He and a friend had slid into a crack in the seafloor and had found themselves in a plummeting tunnel; they had descended to a dangerous depth when their main light went out. Improvising, they used the electronic flash on their camera to peer for a second into the abyss. Only when the picture was developed did they see that the crypt below contained stalactites! These formations can form only in dry air—meaning that the water level of the Caribbean had changed, that the chambers twining below the floor of the Atlantic once tunneled through hills above the ocean. We wanted to reach this extraordinary cavern; we spoke with excitement about the prospect of documenting it fully on film.

Any underwater cave exploration poses forbidding risks. In an emergency in open water, a diver can head for the surface directly above; in an undersea cave, his only exit is all the way back at the entrance. In this sense, a cave sentences a diver to success—or condemns him to death. Underwater labyrinths have confused and ultimately claimed the lives of many divers. François de Roubaix, a talented young composer whose music I considered for some of my films, drowned in a cave off the Canary Islands. My friend Conrad Limbaugh, an excellent diver, was bedeviled by the underground river of Port Miou, France. Bébert Falco joined the group that went searching for Limbaugh in the flooded maze and found his body. Limbaugh had undoubtedly met a horrific death, coming only after the desperate realization he'd lost his way.

The Bahamian caves pose particular hazards. Benjamin reported they had taken the lives of more than twenty divers during the single year before our mission. The hundreds of cave entrances that dot the ocean floor around the Bahamas lead to a webwork of tunnels, making the entire sea bottom seem like a giant calcified sponge. Outflowing tides create such surging currents within these intersecting corridors that waters actually spew and bubble as they gush out the entrances, boiling up for six hours at a time. Inflowing tides, in turn, spin down the cavern mouth in forceful whirlpools, whose hellish sucking sounds seem to resound with the warning that divers who are careless will be swept into the belly of the earth. If we hit upon a plan to outwit these perils, moreover, we would have to enact it with minds dulled by nitrogen narcosis; Benjamin's chamber beckoned just at the level where Rapture begins her deadly and seductive flirtation with the human brain.

For a dive that would last minutes, we thus prepared for days. In order to measure currents around the clock, we installed instruments in the opening of the cave, a narrow slot in the ocean floor, just ten feet below the surface. We identified an interval between tides—which lasted only twenty minutes—in which we could safely dive into the crevice. On each of the next few days, small teams took advantage of these slackwaters to prepare for our filming day. The first afternoon a group of divers began to unreel a heavy guideline to mark the way into and out of the cave's convoluted corridors. They inspected the cave's entry portion: a slanting, chimneylike chute that dropped 165 feet down. The following day a team continued to unreel the guideline as they explored the next section of the cave, a horizontal tunnel that stretched for eight hundred feet and was partially blocked by landslides. Finally, a team, always unreeling line, followed this horizontal corridor almost to its end, leaving extra air tanks near the entry to Benjamin's chamber itself.

With our path fully mapped, we knew that we could not swim fast enough to reach the chamber, then film it, in the short interval of dead calm between tides. We would have to turn the tidal currents to our benefit, to use the faint endings and beginnings of the tides to pull us toward the cavern, then propel us out. We allotted ourselves exactly ten minutes to reach the chamber, five to film it, and ten to return. We were cutting it close; even a brief delay could cost us our lives.

The day of filming, we stood, fully equipped and poised together in our Zodiac, tensely counting down to starting time. I gave the signal for "Go!" Each of us individually slipped headfirst through the slot that opened in the ocean floor. Our tanks scraped and clanged against the narrow chimney shaft as we dropped through the 165-foot chute. We could not afford a moment of claustrophobia; we had to press on in our race against the tide. By the time we reached the horizontal passageway, two minutes had passed. Eight remained to swim the eight-hundred-foot corridor. We pushed ahead.

We kept the rhythm of our fin strokes under control as we progressed into an eerie world, dispersing eons of darkness with the dancing beams of our flashlights. With 165 feet of solid rock above us and the constricted, rock-strewn tunnel ahead, we struggled to fight off alternating waves of foggy Rapture and inevitable apprehension. The trick was to accept the fear—to think, "I'm afraid; so what?" Once you're in a cave, you're in. There's no doubling back along a line of divers in a narrow one-man passageway. Our eight minutes for traversing the corridor passed. We had followed the tunnel to its end.

Upon a signal from the cameraman, each of us ignited a high-intensity magnesium torch. A vaulted cathedral suddenly opened up before us. The torches cast unworldly shadows of fire in water, flickering on the walls of the chamber, revealing the towering arches of its ceiling, which loomed fifty feet overhead. Its cloistered walls echoed with a cantata rumbled by the flames from our torches. The entire dome above dripped with stone—stalactites! They inched down almost to touch the stalagmites that had grown up under each, vertical rivulets of rock, forged by time, testaments to an era when bats, instead of fish, may have sought refuge here.

Reality startled us out of our reverie when the rumbling stopped. Our flares had burned out, filling the cave with smoke. We remained in the confusion for an endless moment; then we followed the guideline like Ariadne's thread, back through the tunnel, to the chimney, and up its chute. We reached our first decompression stop, twenty feet below the entrance, just in time: The outgoing tide began to belch its angry waters upward. Holding tight to the rope, we fluttered in the gushing spout like *Calypso*'s signal flags whipping in a storm. Now we felt no fear—only elation, at having accomplished the most thrilling dive of our lives.

The opposite end of risk—when elaborate precautions collapse like a house of cards—tests mind and mettle far more severely. The lives such risks have claimed testify not to failure, just to fate. The ice-choked Antarctic seas brought just such adversity to *Calypso*'s team.

We did not set out for Antarctica naïvely. We knew its remorseless winters had taken many lives and wrecked dozens of renowned ships. The *Belgica*—with the young Amundsen aboard—was trapped by a frozen sea for the entire winter; three of the men went insane and another, dead and buried at sea, haunted the forlorn group beneath the ice at their feet through the long Antarctic night. The *Antarctic* was lost at the end of February, the beginning of the winter season, in waters not even as near to the South Pole as I intended to sail. Ernest Shackleton's *Endurance*, frozen into the Weddell Sea, became a virtual ghost ship, motionlessly sailing a field of ice until floes broke and crushed the wooden vessel like a walnut.

Calypso, with its hull of wood, was not an ideal ship to sail to Antarctica. But *Calypso* was the only ship I had. And the reasons for accepting the Antarctic's challenge were compelling. Whales had been hunted almost to extinction, and I needed to understand their plight better if I was to fight for their protection. Diving had been rarely used to study the extremely distinctive marine fauna of Antarctic seas, and no exploration submarine had yet probed the deep waters around the frozen continent. NASA sweetened the enticements by requesting that we gather basic data on surface chlorophyll and oceanic currents. The space agency also provided *Calypso* with instruments that gave us direct access to three orbiting satellites. The first ship ever outfitted with such equipment, *Calypso* could monitor meteorological information and receive photographs of the extension of the ice pack as we navigated.

During December and January, the Antarctic revealed even more marvels than we had imagined. We deciphered the history of volcanic eruptions, engraved in the ice walls of a fractured glacier. Under the ice we listened to the choirs of Weddell seals. We dug out fossils of plants and trees torn from India by continental drift. We saw creatures, clinging to deep-water cliffs, extend an endless sticky tentacle to ensnare shrimp and retract them into a shapeless mouth. We roamed in whale cemeteries scattered across liquid prairies. And we watched with an almost religious fervor as penguins performed their ritual of procreation, unperturbed under the volplaning snowflakes and the dive-bomb attacks of skuas.

On February 12, as we approached our scheduled date of departure, we spent a glorious morning. We planned to leave this desolate but dazzling land in a week, well before winter weather was expected to begin at the end of the month. We arrived in Hope Bay, in the Weddell Sea at the northwest tip of the continent, on glassy seas beneath clear skies. My son Philippe and his team took advantage of the placid weather to dive along a giant iceberg. Enticed by the splendor of the glittering mountain, they slipped into a crystal cave, penetrating the iceberg itself, where they filmed the mirrored nooks and crannies inside the undersea ice castle. The constant satellite data we received thanks to NASA indicated that a depression would bring a storm the next day. I had no concern for the moment; wind velocity at the time was at zero. But nature had deceived the satellites. The divers were just returning from their escapade in the ice castle when snow began to fall, big, thick, wet flakes that reduced visibility to fifty yards.

The storm hit literally in seconds—the same kind of sudden tempest that had splintered the historic exploration ships and stranded their crews out on the floes. We had no time even to reel in our anchor; a fusillade of windswept ice blocks blasted toward us, chunk after chunk, bombarding the ship. So fast and so repeatedly did the floes smash us that we neither saw nor expected the mammoth iceberg, ten times the weight of *Calypso*, that careened into our hull, bashed a hole two feet above the waterline, and slammed our starboard propeller—one of only two propellers on the ship—badly bending it. *Calypso* shuddered as she absorbed the blow, and then the winds raged on, pushing the onslaught of ice blocks out of the mouth of the bay. Only minutes had elapsed. The wind had erupted from zero to sixty-five knots—dead calm to whole gale force. I realized we were about to face an Antarctic blizzard.

Our anchor held us in the icefloes' path like a fiendish hand from below, gripping a swimmer's ankle in a nightmare. We could not remain moored, even inside Hope Bay. A massive glacier towered precariously over the far end of the bay. We had seen these monumental walls collapse of their own weight in calm weather, colossal mountains crumbling before our eyes. If the cliff could not sustain the ruthless winds, if it calved, we would have to move out of the trajectory of its crashing blocks. But neither could we leave the shelter of the bay; in the open sea the situation would be even worse: There, the cohorts of gigantic icebergs would tear us apart in minutes. We could only haul up our anchor and dodge ice as we stayed in the relative protection of the inlet. Some lost soul had not named it Hope Bay for nothing.

Three of us split the grueling job of round-the-clock watches: *Calypso*'s master, Alain Bougaran; my friend Roger Brenot, a fellow Navy officer and a contemporary; and me. We attempted to keep *Calypso*, limping on its one good propeller and one bent one, sufficiently maneuverable in the narrow bay in case the glacier foundered and its icy mass exploded like shrapnel.

In just one hour the winds accelerated to eighty-five knots, hurricane velocity, with peaks of one hundred knots. Now we were blinded. We could not even see the bow of the ship, just snow flying horizontally in the wind, turning day to dark. Our ordeal had only just begun. I spent the next twenty-two hours on the bridge, interrupted only by a hurried dash to the mess to warm up with a swallow of coffee. No one remembers when the first day ended and the second began.

I was hastily pulling on some dry clothes in my cabin, during Brenot's watch, when I heard a reverberating *CCLUNGGG*. I phoned the bridge. Brenot answered tersely. He was revving up the port engine, but the unbent port propeller didn't respond, and the ship wasn't turning. The port shaft, the cylindrical rotating rod that connects the propeller to the engine, must have broken. Now we had to rely only on the starboard engine, connected to the badly bent starboard propeller. We could turn in just one direction. We were reduced to facing the furious storm by going in a circle.

Our lives depended on radar and on the towering ice cliff's resistance to the winds. If our radar went out in the blinding blizzard, we would be unable to see the coastlines, and without a doubt *Calypso* would run aground and sink. If the cliff collapsed, the resulting tidal wave and hurtling ice blocks would wreck the ship; we would have time only to send a probably hopeless SOS before piling onto rubber rafts and surrendering to heaving, half-frozen waters where no rescue operation could ever find us. No one on board spoke of the dangers. No one even thought about them. The mechanics desperately watched dials and attended to the least detail in the hope of keeping engines running; my wife, Simone, tirelessly made gallons of coffee, piles of sandwiches, and traded warm gloves for the soggy ones she would dry on the stove. The second night elapsed into the third morning.

On the third afternoon, the blizzard abated. *Calypso* wobbled, top-heavy under thirty tons of ice. Not knowing how long the calm would last, we hurried to hack away the ice and temporarily repair the hole above the waterline in our hull. We alerted NASA that we would take advantage of the lull to make a run for it, from Hope Bay around Antarctica's northern tip and up to a better anchorage at King George Island. We made it.

Yet we still had to undertake the next dangerous leg of the journey, from King George Island across the notoriously stormy Drake Passage. With only our starboard shaft intact, connected to a bent propeller, we had become vulnerable; if this shaft broke, we would be stranded, crippled in the passage known for some of the roughest weather on the planet. We arranged for an escort, the Chilean naval vessel the *Yelcho*, to take us in tow in case of such an accident. The Antarctic gave us a parting break. The notorious narrows remained flat as a tabletop. We safely reached South America and dry-docked *Calypso* in Punta Arenas to examine the quickworks of our battered ship. Just as we had expected, the portside shaft, connecting the engine to the propeller, was smashed in two. What we hadn't expected was that on the other, starboard, shaft, four of the five bolts had disappeared. We had weathered the misadventure on one bent propeller hanging by a single bolt. Had that bolt given way in the blizzard, my men, Simone, and I—we all would have lost our lives.

Had I to live the episode again, I would repeat it. Exactly. We had made every preparation possible. We had applied reasonable means to a reasonable goal. My men, Simone, and I—when we mentioned those days, we did not waste our time speaking about how we nearly perished. We spoke of the ways we managed to survive.

"The utility of living consists not in the length of days but in the use of time," Montaigne wrote. "A man may have lived long, and yet lived little." The most dangerous room in the home is the bedroom, the principal scene of the most common fatal home accidents—deaths by falls, fires, poisoning, and suffocation. Another perilous place is the bathroom, where hundreds drown in tubs. There must be a better way to die.

Surely there is a better way to live.

CHAPTER THREE

PUBLIC RISK

D AY after day, newspapers publish new versions of the same story—another chemical spill is seeping into a neighborhood in which the pediatric death rate simultaneously rises; another leak of radioactive waste is filtering into groundwaters; another new technology threatens to disrupt global climate patterns. And day after day, authorities issue reassurances to stave off what they characterize as the public's frivolous fears. Experts, they say, deem the hazards acceptable, for reasons too complicated for average people to understand; human beings must accept risk if the human community is to progress.

No one can deny that we must accept certain justified risks in order to build a dynamic civilization. Yet throughout history, humans have strived to create a society that is not only dynamic but also shaped by democratic ideals. Liberty, fraternity, the freedom to make personal choices, and the right to full information—these are the very values compromised when officials prod members of the public into taking haphazard risks without careful preassessment, often without even their knowledge.

One day, just as we were completing an expedition, we received news about a catastrophe involving five volunteers whose heroic risk had gone horribly wrong. Inevitably, our thoughts fell to the dangers we ourselves had just faced and the reasons why we had chosen to face them. That one sad afternoon defined for us the glaring difference between individual risk that has been personally accepted and public risk that has been impersonally imposed.

For years my friends and I had dreamed of designing a windship—a boat that would conserve limited fossil fuels, polluting neither air nor ocean, but that could match the speed of modern engine-driven vessels. Inventors had already proposed innumerable systems to succeed canvas sails. One—a towering, upright cylinder that rotated at high speed as it transformed wind into propulsive thrust—overpowered all its competitors, but the huge spinning structure posed serious dangers on board a ship. We decided to aim for the impossible: We would try to invent a new windship system using a cylinder that did not spin but that could nonetheless exploit the wind as effectively as the rotating one.

In conjunction with Professor Lucien Malavard and his brilliant doctoral student Bertrand Charrier, I experimented with countless refinements in some six hundred wind-tunnel tests and finally produced what we called the Turbosail. Like the old wind system, the Turbosail is a hollow cylinder that stands upright on a ship; but because of its specially designed cross-section and aerodynamic additions, such as flaps and a fan, it works without spinning. The Turbosail functions on much the same principle as an airplane wing; instead of exploiting the wind for a lift into the sky, as a horizontal airplane wing does, our upright "wing" exploits the wind for a pull through water—functioning safely, adapting to any weather condition, and operating about six times more efficiently than classic sails.

Eager to play with our new invention, we decided to try an Atlantic crossing. With five enthusiastic companions, I was soon sailing from Marseilles to New York with a secondhand catamaran on which we had hastily erected the Turbosail cylinder.

The first few days proved both the superiority of our new system and the inadequacy of our cheap boat. In the gusting winds, the Turbosail pulled us so swiftly that I sometimes forgot myself and shouted aloud in sheer joy. Meanwhile, the antiquated catamaran jolted incessantly, lurching up, then slamming down with a bone-jarring, tooth-loosening thwack.

The third week, we entered a storm. Battered by even more forceful shocks, the Turbosail began to split off from the welded seams that fastened it to its base. If the welding gave way completely, the lofty mastlike structure would topple over and smash onto our deck, we would be defeated by that miserable, cheap, bounding boat, and our failure to reach New York would reflect unjustly but inevitably on our now-proven invention.

To avoid disaster at least temporarily, we braced the tower with a kind of internal rigging, crisscrossing tight steel wires inside the hollow column to fasten it to its base. We managed to reach Bermuda and asked shipyard workers to re-weld the split seams.

Then we headed back to sea.

The very next day even more savage winds arose. Before long, we were fighting a gale. The freshly welded seams split. Again we braced the Turbosail with a tightly strung web of wire. Just as I assumed the midnight watch, I heard, over the howling wind, a sound like a cracking whip. One of the steel wires had snapped. I grabbed a flashlight and pointed its beam across the wave-lashed deck. The split at the cylinder's base had widened and now gaped menacingly. If the giant structure crashed onto the boat, it could kill us. I seized the wheel and veered to put the wind behind us just as the majestic tower capsized over the bow and was instantly swallowed by the sea.

The accident had spared our lives; but it had left the six of us on a boat with a yawning hole in its deck, tossing relentlessly in a gale-battered sea. Our radar equipment, also shorn off by the wind, was now sinking with the cylinder. I climbed down into the mess, where my five colleagues had gathered in momentary shock. Our windship was ruined. We agreed to start over.

We rode out the storm, returned to France, planned, designed, and built an entirely new ship, *Alcyone*, and once again set out across the Atlantic. No used catamaran she, *Alcyone* had a slender prow that sliced the waves; she was outfitted with two of our Turbosail cylinders, both of them reinforced in such a way that even a hurricane could not snap them off. Using only half the fuel of a comparably sized petroleum-burning ship, we sailed from France to a fireboat welcome at New York harbor. Then we pressed on. We didn't want to know simply if *Alcyone* worked; we wanted to know just how much she could endure.

Mount Horn looms from the last forsaken island off the bottom tip of South America, where the Andes spill their rocky treachery into the sea. Thousands of ships have been driven ashore and splintered by Cape Horn's terrifying gales; even some captains who have saved their boats have lost their crews when winds in these "Roaring Forties" latitudes swept men right off decks and into the angry ocean. Here *Alcyone* proved her mettle for two months.

Exhilarated and vindicated by our hard-won success, we wanted to share our excitement about *Alcyone* with the Horn's only residents, and so one afternoon, we scrambled halfway up the mountain to a bleak cabin where two solitary Chilean Navy sentinels manned the island's lonely radar post.

But that particular afternoon, that particular hour, a celebration was not to be. As we arrived, the two sentinels, their faces ashen, stepped outside their cabin. Their radio was broadcasting shocking news. The American space shuttle *Challenger* had just exploded.

Simultaneously, people around the world experienced the stupefying blow of extreme emotions reversed, of the thrill of a rocket launch turned to grief, of hope turned to horror. Up on our desolate, storm-swept crag—where the wail of the wind always seems to lament sailors who perished at the Horn—we felt especially haunted by the dual face of exploration: its promise and its price. We realized that our own pre-occupation with inventing the windship, with outwitting the ocean's elements, had taken our minds off our expedition's risks. On the rocks below, fuming waves marked the grave of wayfarers who in centuries past had died in efforts to enlarge the human scope of seven seas. Above us only the empty sky memorialized the great goal undertaken—the greatest sacrifice made—by the astronauts who had tried to enlarge the human scope of heaven.

Those pioneers of sea and space had fully recognized the risks they faced. They had accepted them. They had judged the potential risk to their own lives as outweighed by the potential benefit for humanity. The astronauts had not ridden astride their veritable hydrogen bomb for the thrill of tempting death. They had risked dying in their zeal for living, for fulfilling dreams, for enlightening generations to come.

Progress is indeed bought with risk: In that respect, authorities who goad the public into risk taking are right. Where the authorities go wrong is implying any connection whatever between risks that courageous people face *for* the human community and unreasonable risks that technicians impose *on* the human community.

Unreasonable public risks are not undertaken to achieve civilization's ultimate goal—to preserve and promote the sanctity of human life. Cutting corners on nuclear-plant safety, disregarding health hazards represented by certain lucrative chemical products—such risks display no reverence for life, just indifference to it. They do not aim for selfless humanitarian goals; they aim for simple profit. Inflicted by those who reap the short-term benefits, these risks are faced by citizens who sustain the long-term costs.

Unreasonable public risks are not accepted by volunteers. Too often public risks are hidden from the public, censored by governments and industries that illogically justify their endangerment of human interests by citing the national interest.

Unreasonable public risks are not carefully planned endeavors in which all predictable hazards have been painstakingly identified and systematically eliminated. Those of us who plot each detail of a mission know the price of risk. We recognized the anguish in the faces of the astronauts' parents as they gazed at the *Challenger*'s broken trail of smoke. I saw that same expression of pain when I told the parents of one of my team members that their son was dead—killed in a freak accident during our voyage to the Antarctic. I bore that unbearable grief when I learned that my own son had lost his life in an unpredictable accident during a mission. We who have paid the price of unforeseeable risk—who thus take excruciating precautions against foreseeable risks—we listen in astonishment as technicians exhort the public to plunge into dubious technologies, to rely on blind faith as a guide into the future.

Many people rest assured that because government has approved a product, it must be safe. They dismiss new media revelations about dangers improperly imposed in the past as irrelevant to their own present lives. Emerging from the ever-growing pile of newspaper clippings, however, is something far more significant, far more disquieting and insidious, than seemingly unrelated anecdotes about sporadic public dilemmas. The reports show a pattern, a matter of input and output. If long-ago risks are being revealed every day, logic demands that we assume that risks were—and continue to be—imposed day by day.

Technocrats are turning us into daredevils. The haphazard gambles they are imposing on us too often jeopardize our safety for goals that do not advance the human cause but undermine it. By staking our lives on their schemes, decision makers are not meeting the mandate of a democratic society; they are betraying it. They are not ennobling us; they are victimizing us. And, in acquiescing to risks that have resulted in irreversible damage to the environment, we ourselves are not only forfeiting our own rights as citizens. We are, in turn, victimizing the ultimate nonvolunteers: the defenseless, voiceless—voteless—children of the future.

If the modern mismanagement of risk was overseen only by corrupt politicians and evil technicians, the story would be more melodramatic to tell and the problem simpler to solve. Instead, mistakes in risk management arise from the fact that, as technology advances, we are losing sight of just where we want it to go.

Certainly, the challenges we've faced along technology's path have been daunting: Assessing the safety of new products, neither understating danger to society nor needlessly denying the public a beneficial advance, is no easy task. Even as notorious a toxic pesticide as DDT presents the public with complicated choices. DDT acquired its dubious reputation after scientists discovered that the pesticide had killed birds, fish, frogs; that winds and ocean currents had dispersed DDT around the globe, with levels appearing even in the Antarctic snow; that it had concentrated in the milk of mammals; and finally, that whole national populations had been exposed to the point that an appalling generalization could justifiably be made: The average citizen carried tissue levels of DDT. But DDT had also successfully achieved the purpose for which it had been introduced—it had arrested the spread of typhus by body lice, malaria by mosquitoes, typhoid by flies. Sri Lanka's ban of DDT was promptly followed by an explosive malaria outbreak. Poison or plague: which to choose?

And so the field of "risk management" was born of a good idea. To resolve the moral quandaries presented by modern products, experts would analyze technological innovations, calculating the severity of their possible hazardous effects and the extent of their benefits, to provide solid information for decision makers who accept risks on behalf of the public.

The intent is admirable; the underlying premise, fatally flawed. Calculations are only as "solid" as the figures fed into them. Risk-benefit analyses, based on unfounded assumptions and debatable statistics—predicting unpredictables and quantifying invaluables—can simply never supply decision makers with solid information.

Risk assessors' first mistake involves the basic assumption they use as a foundation for all their subsequent conclusions: that human gains and losses—the "benefits" and "costs" in their equations—can be represented by something as one-dimensional as a statistic. How can anyone—mathematician, statistician, statesman—ever succeed in providing a number to represent intangibles like good health, peace of mind, joy, or the anguish of disease, premature death, loss of a parent, spouse, child?

Risk statisticians have unfortunately met this challenge by proceeding as though it does not exist. In evaluating the benefits of a particular risk, they simply exclude educational, environmental, philosophical, or charitable gains. Corporate profits are quantifiable; hence profit becomes the only benefit that literally counts. Worse, loss is routinely assessed in cash as well.

I still feel rage when I recall a meeting during which a "risk evaluator" reviewed the community's losses following the infamous 1969 oil spill off Santa Barbara, California. As he began to tote up the environmental damage, he made an offhand comment: "We'll appraise each dead bird at a dollar apiece, as a matter of principle."

I was so stunned I blurted out, "Principle! How dare you put a cash value on a bird? On life?"

"We have to put a figure here."

I asked him rhetorically if he would put a price on my grandchildren. In fact, as a risk assessor, he probably already had. Many of us are scandalized by lands that compensate a murder by assigning a legal "price of blood" according to social status. Ironically, supposedly civilized countries routinely assign a cash value to life according to the financial status of the deceased—based on projected losses to the economy of the individual's earnings. The U.S. White House, for one, once argued that since those exposed to asbestos may live some forty years after exposure, risk assessors should "discount" the losses incurred with death, i.e., the inherent value of a human being, when considering laws for asbestos-based products—marking the price they originally assigned to life, of $1 million, down to $22,094.98. Another cost-benefit analysis, conducted by Brown University and Rhode Island Hospital, concluded that the cost of saving a low-weight, premature baby exceeds the child's potential lifetime earnings and that rescuing heavier babies would return more than half the expenditure to society. Even the most elementary critical thought reveals the absurdity of this rationale. Who believes that an arms dealer—with his exponential earnings—has a higher inherent human value than a Mother Teresa? Or a Gandhi? Or a Mozart?

Consider a single scientific achievement: Not long ago, researchers tried to replicate the ingredients of the primal seas in a laboratory broth. They then exposed the concoction to electrical charges, hoping to mimic the effect of the newborn Earth's lightning. The result: Uric acid appeared in the laboratory flask. The accomplishment produced worldwide astonishment. The miracle of life remains so far beyond our comprehension that even the laboratory production of uric acid—life's waste product—prompts global celebration.

Life cannot be assigned a cash value because, simply, it is beyond value.

Yet the Golden Calf syndrome—assigning great value to metal and no value to life—persists in risk management, infecting not just theory but also practice. The record is incontestable. Again and again, market values, rather than human values, have dictated policy decisions:

- Two weeks after thousands died in the 1984 disaster at the Bhopal chemical plant, the United States dissented from a U.N. accord to publish a directory of hazardous substances—a simple list that would merely have identified dangerous products. Why? Publication "could unfairly discriminate against the export and sale of products of certain companies."

- In nuclear-power analysis, the spurious belief that life can be priced led directly to the spurious conclusion that if an "extreme catastrophe" were to wrack the same plant every three years, deaths would be "acceptable"; "economic consequences," however, would be "completely unacceptable." The numbers inspired technicians' call for stricter safety regulations, not to protect lives, but to protect investments.

- In the mid-eighties, when West Germany, its forests devastated by acid rain, proposed a date for limiting car-exhaust emissions in Europe, neighboring nations angrily denounced the idea and successfully postponed the deadline. No one denied that acid rain reduces food harvests, precipitates the release of poisonous heavy metals into groundwater, even threatens hundreds of thousands of Alpine villages with avalanches that the withered trees can no longer restrain. Italy, France, and Britain instead protested that the cost of installing the necessary catalytic converters would reduce auto sales. The decision to postpone protection of the quality of Europe's air and the quality of European life was made for the welfare of the European auto industry.

- When announcing their decisions to join the 1980s U.S. Star Wars program to orbit weapons in space, not one participating nation addressed physicists' objections that the plan was "deeply misguided and dangerous," in the words of one Nobel laureate, that the arms might not work, that the weapons could endanger the very people they were meant to protect. Instead, every single nation based its decision to place weapons in space on financial projections: Britain joined the program with the announcement that the U.S. secretary of defense had promised "a very substantial" number of contracts to British arms manufacturers. Japan joined with the declaration that the program would yield technological spin-offs for Japanese industry.

France not only joined the U.S. program but also proposed a space-weapons program of its own. In the United States, proponents of the program argued that *even if the weapons plan backfired*, as physicists forewarned, the estimated $20 trillion for research would forge a "silver lining" for the arms industry. "What else would we do with the money anyway?" queried the president of one consultancy firm. "Give it to welfare, to health care, to Social Security? These are all sinks. They do not reproduce . . . create new things." The remark gives new life to the old joke about a miser's response to a gun-wielding robber: "Take my life. I need my money."

Another practice that renders risk assessments invalid involves the mathematical sleight of hand that risk assessors use to transform their calculations, manipulating the numbers to produce veritably any conclusion they want. In my experience with exploration, calculating an acceptable risk has meant eliminating peril. In official risk management, it means eliminating the *appearance* of peril.

Juggling the figures obviously defeats the purpose of risk assessment; the amoral way in which figures are juggled defeats the ideals of democracy as well. The most stunning void of ethics surrounds the use of a tool known in risk parlance as the "discount." Statisticians have, for example, calculated the projected loss of human life from pollution, overpopulation, and exhaustion of limited resources. When they concluded without dispute that the resulting death toll will be "utterly unacceptable" for the next generation, what was their remedy? They reasoned that the public worries little about risks that cause death in the distant future or in distant lands, and so they applied their notorious "discount"—with which they reduced the cash value of the pertinent deaths. The statisticians didn't argue whether disease and death will matter to victims in the distant future or in distant, developing nations. They simply contended that those diseases and deaths don't matter to us here and now.

Practices like these have prompted at least one critic to dub risk assessment as "mathematics masquerading as science." Yet risk assessors doctor the accounts in other ways as well. Once they calculate their odds that a given technology will cause a catastrophe—itself an exercise in conjecture—they face a judgment call: Just what odds of disaster are acceptable? Thus statisticians have chosen various reference points against which to measure a given chance of catastrophe. Some risk assessors submit, for instance, that a technological risk is "acceptable" if its odds of ending in tragedy are at least ten times lower than the odds of a natural disaster. "Examples," writes one government adviser, "are the probability that the sun may explode in one's lifetime and the risk of being hit by a meteor falling to earth . . . These risks are unavoidable and uncontrollable, at least with present technology; therefore they have been accepted by society and are generally ignored."

This logic is comforting but counterfeit. Who has "accepted" death by meteorite? People do not *accept* "acts of God"; they *endure* them. When one government adviser minimized the possibility of an atomic-power catastrophe by asserting that "earthquakes, hurricanes, and tornadoes are much more likely to occur and can have consequences comparable to or larger than a nuclear accident," he omitted mention of the simple truth that undermines his argument: We can't prevent earthquakes, hurricanes, and tornadoes; we *can* prevent unnecessary technical disasters. One nature does to us; the other we do to ourselves.

Those risk assessors who recognize the inadvisability of comparing acts of men to acts of God suggest an alternate but equally specious comparison. A technological risk can be acceptably imposed on the public, they have concluded, if its odds of disaster are no worse than those of risks that government and industry have inflicted on the public in the past. People, as one specialist explained, have already come to regard past hazards with "resignation and reluctant acceptance. The acceptance level is a reference against which new risk is determined and then compared."

According to this scenario, we become "resigned" to a risk imposed without our consent; we are deemed to have "reluctantly" accepted it; standards go down; resistance goes down—how can progress possibly go up? The rationale hardly supports risk managers' own grandiose contention that humanity must take risks to improve the human condition. They do not aim for the highest safety criteria human ingenuity can conceive—just the lowest that human beings can bear.

Still another way that risk calculators manipulate their conclusions about risk—and thereby manipulate the community that faces risk—is by feeding oversimplifications into the formulas. If statisticians don't know how to create a computer code that accurately represents a complicated but disastrous possibility, they simplify it to suit their software or ignore it altogether. Thus, around the world, nuclear-plant safety projections have variously omitted the possibilities of accidents caused by flood, earthquake, airplane crashes, and sabotage. In considering one nuclear power plant built three miles from an earthquake fault, for example, risk assessors dutifully calculated that the chance of an earthquake's occurring at the same time as a nuclear accident was just 1 in 6.5 million; inexplicably absent from their calculations—and their subsequent safety assurances—was the chance that an earthquake could *cause* a nuclear accident. Another tortuous computation of a serious accident left out so many factors that mathematicians could only conclude that the chances of catastrophe fell somewhere between "the impossible" and "the inevitable."

Far more disquieting than statisticians' deliberate omissions are their inadvertent omissions, the possibilities they exclude from accident projections, not because they don't know how to calculate them, but because they have yet even to imagine them—they call them the "unknown unknowns." Industry commonly uses tens of thousands of chemicals, for example; few people realize that only an estimated 20 percent of the products in everyday use have been tested for health effects. With no inkling about the individual effects of thousands of individual chemicals, how can anyone predict the powers they may wield when mixed together, in limitless combinations throughout open air and water where we spray them, spill them, dump them? Already some communities have no money to pay for, or know-how to remedy, the effects of toxic wastes and radioactive spills from accidents to which officials attached little importance in years gone by. Their future is our present; who were these officials whose only concern about danger was that it affect us and not them? Do we really care so little for our children that we too can dismiss the unknown costs of untried technologies in an unimaginable future?

It could be argued that even uncertain figures could project at least a tentative light into the darkness ahead. True—if decision makers presented their safety and risk projections honestly, as ambiguities rather than as absolutes. Instead authorities brandish the estimates as incontestable truths, announcing urbi et orbi that technologies are unquestionably safe. The record has too often shown them to be unquestionably wrong:

- "For catastrophic types of nuclear-plant disasters, one type—core melt—would occur on the average of one every 17,000 years per plant," the Rasmussen Report, for years the nuclear-risk bible, guaranteed in 1975. By the time all the world's nuclear plants had together operated a total of only 4,000 years, the world had already sustained two core melts.

- After the U.S. Three Mile Island nuclear accident, Soviet authorities projected that such an event could never happen in a Communist country; the disregard for safety at the American plant, they asserted, typified capitalism. "Soviet safety norms rule out an escape of radiation," the authors flatly stated—seven years and three weeks before Chernobyl irradiated 75 million people in the Soviet Union alone and, some experts believe, helped to topple the Soviet empire and to end the Cold War.

- We who were so deeply affected when we learned of the *Challenger* disaster can feel only sorrow when we read the definitive safety guarantee that had been attached to the fated shuttle. Rocket technicians specified that their space shuttle would "fail" once in 100,000 firings. The *Challenger* exploded on the twenty-fifth firing.

Such utter implosions of safety guarantees shock the world but fail to humble the statisticians. Policy makers march unflinchingly onward, oblivious to the mistakes of the past, issuing more unequivocal guarantees for the future. In legend, Dionysius, tyrant of Syracuse, purposely undermined his peoples' sense of safety and security by tying a naked sword to a horsehair and suspending it above the head of Damocles. We sit beneath Damocles swords that our leaders seem to manufacture by chain production:

- As evidenced by the crash of two U.S. nuclear satellites, as well as Cosmos 950, 954, 1714, and 1900, the record of orbiting atomic reactors has been grim. Despite the fact that one tenth of all Soviet and American reactors launched into space have ruptured, pouring out their radioactive contaminants, policy makers decided to launch a double thermonuclear generator as the first radioactive package to be carried on a space shuttle. The original launch date they scheduled turned out, in retrospect, to be chilling. They slated takeoff of the massive nuclear package for the never-to-be next flight of the *Challenger*. They failed to be humbled either by the *Challenger* disaster or by the alarming result of their own tests. Officials continued to hold to the astonishing contention that even if a reactor had been aboard, it would "certainly" have resisted rupture—despite the fact that their own lifelike mock-up reactor not only exploded but also showered debris, blasted into dust, over a 75-by-200-mile expanse. They furthermore continued to issue reassurances that they could safely launch nuclear reactors on their other space shuttles.

- Add to the Damocles sword of the orbiting atomic reactors a threat that specialists, with their tunnel vision, have considered separately: Space is littered with hundreds of thousands of pieces of debris, the refuse of satellite programs conducted by China, Japan, India, West Germany, France, the European Space Agency, Britain, the former Soviet Union, and the United States. These pieces of garbage— which can stay in space for three hundred years—orbit Earth at more than Mach 20, ten times faster than the Concorde. A collision with a marble-sized metal chip flying at this speed will devastate a satellite; flying fragments from a shattered satellite could in turn set off a cascade of more destruction. Have decision makers considered the risk of a crash between their orbiting shrapnel and their orbiting reactors? After years of producing debris and launching reactors, officials conceded to my coauthor on this book that, aside from considering the problem a real one, they otherwise hadn't much considered the problem at all.

Such lapses by those who lead nations bewilder explorers who have led a team. During the course of my own career, I learned the prerequisites of risk management after our volunteer Jean-Pierre Servanti had died.

As described in the previous chapter, Servanti died while diving too deep, too long, searching for the anchor we had lost as we excavated a Greek shipwreck off Grand Congloué. Amid the anguish, amid the remorse and mourning, the fact remained that the anchor still lay lost on the bottom. We dragged the area with a sharp grapple and snagged the anchor's chain 240 feet down.

The treachery of the dive had killed Servanti the day before. Who now was going to dive to tie the anchor to a heavy rope so we could winch it up? Diving that deep, that late in autumn, meant struggling against a brutal current in depths where nitrogen saturates the blood, distorting judgment with Rapture of the Deep. I knew that if I was going to function as a leader for my team, there was only one answer. I had to go myself.

I slid into the cold water, fought my way to the bottom, then saw with dismay that I had descended at the far end of the anchor chain, that the anchor itself lay off in the distance. Hand over hand, I pulled myself along the chain, dragging my already weary legs, focusing my thoughts through a haze of Rapture. I made an effort to remain calm, to conserve as much energy as possible. When I reached the anchor, I tied it to our rope and headed for the surface, stopping in the numbing cold for decompression.

I broke surface to learn that my knot hadn't held, that the men had winched up my rope like an untied ribbon, whipping loosely in the wind. The anchor remained on the bottom. I climbed aboard the ship amid an awkward silence. The men averted their eyes from mine. I told them I would take three hours to clear my body of nitrogen and go down again.

Three hours later, still cold and exhausted, I faced another cold, exhausting dive. Even as I battled currents on my way back to the bottom, knotted the anchor tightly on my rope, surfaced, and watched the men winch it up, I knew that the damned hunk of iron was not worth the risk I had taken.

But I hadn't taken the risk for the anchor. The anchor had become a symbol. I had to show the men who had joined me that while I place great value on our ultimate goals, I place greater value on their lives. I had to prove that if I would ever again allow anyone to take a risk, it would be only after I'd tried it first and identified its pitfalls. Then and only then could I say, "Follow me."

Those who plan public risks do not say, "Follow me." They say, "Trust me." Politicians may rarely be in a position to try technologies for themselves. But they are *always* in a position to demand that risks be fully investigated and that the people who face risks be fully informed. Too often decision makers abdicate this fundamental responsibility of risk management. They do not lead us through truly calculated risks for which they have isolated and then eliminated hazards; they instead goad us into a game of Russian roulette, instructing us to pull technical triggers without telling us if there are bullets in the chambers. This is not leadership. This is not democracy. This is technocratic dictatorship; this is market dictatorship.

Officials have from time to time declassified risk files. They have done so, unfortunately, years after the public has lost interest, years after the victims have lost their lives. The past few decades of news reports have not been reassuring; governments of the people, by the people, and for the people have again and again denied full information to the people at the time they needed it. The average citizen has not participated as a knowledgeable volunteer in community risks; he has no opportunity to accept or refuse danger because he often has no idea he is *in* danger.

○ Certainly people used as guinea pigs in weapons tests were not in a position to provide "informed consent." As reported in 1985 by *New Scientist*, not until twenty-eight years after the fact did the U.S. nuclear agency release a report detailing the risk that officials imposed on uninformed human beings during the Bikini-Enewetok weapons tests. Although U.S. authorities knew hours before an atmospheric blast that prevailing winds were blowing directly toward the people of Rongelap Island, they exploded their bomb anyway. They then withheld warnings as the Rongelapese remained outdoors, bareheaded and bare armed, while fine white radioactive ash fell from the sky all day and into evening, even accumulating like snow. Officials then announced that life on the island was safe for the natives, many of whom showed signs of acute radiation sickness. The dubious "benefit" for which authorities imposed this risk is implicit in a 1958 Brookhaven Laboratory memo that cited the need for "greater knowledge of radiation effects on humans" and concluded that "habitation of people on the island will afford valuable data on human beings."

○ Nor did the British seek fully informed volunteers for their 1950 atmospheric bomb test over Australia. Not until thirty-five years had passed did Australians discover that one of the dirtiest of these explosions, about five times more powerful than agreed between the two governments, had sent a cloud of radioactive dust over large parts of the northeast section of their country.

○ Only by chance did members of Canada's Parliament discover that for forty years their military had withheld—specifically from the Canadian Parliament itself—its accords with the U.S. Pentagon, even regarding consultation prior to use of nuclear weapons. The Canadian military's explanation for censorship: The documents were "too politically sensitive" for *politicians* to consider.

Risk managers might justifiably protest that such aberrations of the military world—in which defending people for some reason requires attacking them—exceed the bounds of risk management as well as the bounds of sanity. Yet the market can proceed just as dispassionately as the military in its imposition and distribution of unreasonable risks. After developed nations pronounced cigarette smoking unsafe, tobacco companies began aggressively marketing their cigarettes in developing nations where information about hazards had not been widely disseminated. After certain birth control devices caused internal damage and sterility in women of developed nations, pharmaceutical companies unloaded the goods in developing nations where women were unaware. After West Germany banned thalidomide for its link to the "seal limb" birth defect, its producer sold the drug, for a harrowing thirteen months more, to unwitting Italians. After the Austrian and German people became educated to the threat posed by their burgeoning nuclear waste, the industries involved negotiated to send the radioactive refuse to China and Egypt.

As hazardous materials wend their way from one uninformed populace to another, officials have found an equally circuitous way of keeping information from the people at home. When a test produces alarming results, authorities often not only withhold the data but also order that the test be performed again—and again and again—effectively creating a welter of delay tactics, blocking safety regulations, and instilling a false sense of security in members of the public. So frequently do government and industry agencies launch studies, withhold publication of negative conclusions, and relaunch their research that even one environmental official lamented the great handicap crippling his own agency: paralysis by analysis. Continual retesting provides a subtle form of secrecy for instances in which the public would not tolerate outright censorship. Procrastination or classification: The result is the same. By denying people their right to information about risks introduced into their daily lives, officials deny people their rights of citizenship.

Evidence of paralysis by analysis appears in the news almost daily, but just one example suffices as damning evidence of deadly procrastination: the reaction of government and industry to one of the most toxic substances known—PCBs, or polychlorinated biphenyls.

○ 1965: The United States belatedly commissions a laboratory study of the PCB dioxin, well after the substance has gained a prominent place on the market as a weed and brush killer and has become a favored defoliant with the military, which is using it abundantly in the Vietnam War.

○ 1966: The U.S. laboratory conducting the study discovers that even small amounts of dioxin greatly increase birth defects in animals. The

U.S. Department of Health, Education, and Welfare refers dioxin back
to the laboratory for three and a half years of further study.

○ 1968: Other lab studies provide more evidence that dioxin causes birth
defects. Studies also reveal that after exposure to a dioxin-related
chemical, hexachlorophene, the brains of laboratory animals actually
developed holes. The chemical, they also report, can be absorbed into
the body if applied on the skin—a horrifying discovery, given the
routine practice in American maternity wards of washing newborns in
hexachlorophene lotions. Although the U.S. Food and Drug
Administration severely restricts hexachlorophene, Western European
companies continue to produce products with it for skin application.

○ 1969: Under pressure from a citizens' activist group, the U.S.
Department of Health, Education, and Welfare finally releases its
three-year-old data about birth defects caused by dioxin weed killers.
A U.S. government commission recommends that dioxin be
"immediately restricted to prevent human exposure." The U.S.
government classifies its own recommendation.

○ 1971: Two more panels report on dioxin's physiological dangers. The
U.S. government asks for more recommendations.

○ 1972: In France, four years after research has shown the dangers of
hexachlorophene when applied to skin, forty-two infants die after the
chemical is accidentally added in excess doses to talcum powder. The
government of France unconditionally declares that "dioxin poses
neither ecological nor human health risks."

○ 1976: A factory explosion envelops the town of Seveso, Italy, in a
cloud of dioxin. Company officials withhold the fact that the Seveso
cloud contains dioxin and maintain their silence even as children play
in the dioxin fog and are later admitted to hospitals in serious
condition. Who owned the factory? A Swiss corporation—owned in

turn by Hoffmann–La Roche, the multinational also involved in the talcum powder accident that had resulted in the deaths of forty-two French infants four years before. Meanwhile, the French government, which a year previously had issued an absolute safety guarantee of dioxin, asks for more study.

The effects of this strategy, to keep people ignorant of the hazards around them, were foreseen as long ago as the time when Western democracies were first being conceived, designed, refined. "A popular government without popular information or the means of acquiring it," wrote James Madison, "is but the prologue to a farce or a tragedy, or perhaps both."

Sadly, too many decision makers around the world demonstrate Madison's point, with the tragic and farcical ways they have forced their will—squelching all public objections should risk information leak out to the people, using ridicule and mockery to discredit members of the community who protest. In one text for risk managers, the author, a government adviser, referred to protesters as "Squawkers," an epithet by which he defined any particular group that finds a risk "distasteful or unacceptable" and "blows it up through dire predictions of consequences based primarily on half truths." Squawkers, this government adviser complained, "place a heavy burden on society" because their charges force "government or other agencies to disprove the claims, or to evaluate the claims rationally."

Squawkers who have made this minimal request—that government evaluate claims rationally—have paid dearly. When Rachel Carson published *The Silent Spring*, defenders of the pesticide faith railed that her masterpiece was a hoax, misinformed, distorted, and fanatical. When the Seveso factory showered the surrounding neighborhoods with dioxin, the company's general director appeared on television to say: "Everybody knows that the Italian people, and especially women, are always complaining; it is known that the Italians are extraordinarily emotional." When the Monsanto company created plastic bottles that were found to leach chemicals with unknown effects into their contents, the company's president scoffed that "the very significant benefits of chemistry are falling victim to an advanced state of 'chemophobia.'" When American senators asked if a nuclear-industry plan could jeopardize neighboring populations, a utility executive actually jeered at the elected officials: "This is a nation of heroes, but unfortunately we have our share of cowards."

If such withering diatribes do not inspire people into taking risk, they certainly humiliate them into it. People so revere the adventurous spirit, they so misconstrue the concept of heroism and so fear being regarded as timorous, that they hardly need to be taunted into unnecessarily facing danger. I first noticed this general acquiescence to needless risk during a 1977 Caribbean expedition. *Calypso* dropped anchor off Saint Pierre. Only seventy-five years before, the famous eruption of the nearby volcano Mount Pelée had wiped out the richest city in the French Caribbean, killing twenty-six hundred inhabitants. Even though I'd been told about the newcomers, I could hardly believe my eyes when I saw that they had rebuilt the town and were living in the shadow of the still-fuming volcanic mountain.

This tendency to submit to risk rather than to take action to avoid it has crossed the bounds of time and place. Farmers of the Ganga plain know that each year the worst floods of India will demolish crops, homes, animals, possibly even their families; yet each year the vast majority returns to rebuild. "Bearing the loss seems the only choice universally accepted among them," reported one observer. "They appear to be prisoners of their experience." Norwegians in mountain villages tread below snow-burdened peaks in daily fear of death by avalanche. Californians persist in centering the heart of their electronic industry on platonic faults so unstable that the region is known as Earth's "ring of fire." Hawaiians live within a mile of the Puna District's volcanic rift zone, disregarding the vapors that constantly steam from vents in the earth. Eleven earthquakes have shaken Turkey's Anatolian fault in thirty years; eleven times Turks have rebuilt along fault lines.

So too does the average citizen become mute before the dangers that government and industries have allowed. When herbicides seeped into the water system of twelve towns south of Milan, concentrating in levels unsafe for human consumption, many townspeople continued to drink the water; one of the villagers explained that his neighbors reasoned, "If I wasn't dead the day before yesterday, I won't be dead tomorrow." He added, "But there's fear." After Chernobyl, when authorities derided inhabitants of severely contaminated regions, belittling their concerns as "radiophobia," many people of Belarus responded not by rising up in anger, but by crumbling into a "Chernobyl syndrome," a debilitating existence bereft of all but resignation and despair.

These views of risk have gone awry. People fear standing up to the authorities who impose danger more than they fear danger itself. Bravado is mistaken for bravery. "Squawkers" are disparaged for blocking the official version of progress, when in fact they aspire to real progress—by pushing technologies to ever safer, ever higher, standards. Nowhere have these scrambled values been more evident than on a sign mounted at the entrance to Chicago's 1933 World's Fair, which took as its theme "A Century of Progress." The placard read:

SCIENCE DISCOVERS
TECHNOLOGY EXECUTES
MAN CONFORMS

Is this the "progress" we want to buy with the coinage of human risk? Are submission and resignation the goals for which we should stake our lives, our children's lives?

"Things are in the saddle and ride mankind," Emerson wrote more than one hundred years ago. A century has passed and all that has changed is the acceleration. The Marx Brothers' method of risk management: The world boasts nearly two hundred leaders. How many of them know where they are leading?

The *New York Times* once published some interesting statistics about this new technocracy—not just about those who manipulate it but also about the citizens who are manipulated. After surveying thousands of adults, the paper reported that only 1 percent of those citizens questioned believed that the executive officers of their government have consistently told the truth.

The astonishing headline that should have bannered these figures remains unpublished. No self-respecting journalist could really consider as news the fact that government officials lie and that the people know it. The surprising—the shocking—fact is that people know their officials lie and they do nothing about it.

The problem of modern democracy is not that the people have lost their power, but that they have lost their appreciation for the extraordinary power they wield. Consider one astonishing truth: Famine has never struck a democracy. Despots can mismanage their people's resources, deplete their people's treasuries, stockpile caches of scarce food supplies for themselves—and they do not have to account for their failure in leadership. Sociologists propose that the most potent weapon against starvation is freedom—the freedom for the public to demand and receive information, the freedom for the public to participate in public affairs.

The final decision about whether to involve the public in any particular hazardous endeavor is not a technical decision or a market decision but a value decision. It should not be made by those manufacturing a dangerous technology, nor by those profiting from it. A decision about accepting risk should be made by those who will take it. The expertise required is expertise in the human condition. It is an expertise possessed by the human community.

When government jeers at "chemophobics," "radiophobics," and public cowardice; when the public reacts in disappointment and chagrin—all are missing the point of risk, the purpose of democracy. The significance that will ultimately be attached to each of our individual lives will not be measured in the problems we will have faced during our time on Earth or even in the stoicism or heroism with which we faced them. It will be measured by what we achieved while overcoming them. This, this alone, is progress.

We do not need another Hume, Pericles, Montesquieu, Jefferson, or Madison to understand the scope of what a single person can achieve in a society governed by a people who insist on getting, and who take action upon receiving, full information. When I picked up a newspaper one morning in 1975, I read a piece about a frightening risk that was menacing people living near Otranto, Italy. The ensuing events would lead me to meet a man who practiced the politics of achievement rather than of expediency.

The article I read reported that a year earlier a freighter, the *Cavtat*, had been rammed by another ship and had settled to the bottom of the Mediterranean. There she remained, a ghostly silhouette sailing the sands of the bottom with her deadly cargo: tetraethyl and tetramethyl lead.

If dispersed by seawater and metabolized in fish consumed by humans, these lead compounds cause insomnia, emotional instability, and hallucination. Severe poisoning—caused by a even drop of the compounds in direct contact with skin—brings convulsions, insanity, coma, and death. More than four hundred drums of the lethal substances lay stacked in the *Cavtat*'s hold, with almost five hundred barrels more trapped under her rigging, lashed onto her deck, and listing freely on the ocean bottom just three miles from Otranto, a fishing and resort town. If one of the drums should leak in the hands of divers trying to retrieve them, if children should play close to a leaking barrel washed up on the beach, doctors would be relatively helpless.

I soon learned that those involved with the *Cavtat* had wrapped themselves in a protective tangle of international complications. Rammed by a ship with a Panamanian flag, the *Cavtat* herself was owned by Yugoslavs, who claimed that their cargo was "only" a pollutant, not a hindrance to navigation and therefore not in violation of law. Italian authorities contended that the ship was not their responsibility because it sank just far enough off their coast to rest in international waters. International authorities could do nothing because the pertinent codes had not been ratified by the necessary number of nations.

And what of the little town of Otranto just three miles away? "We reassured executives of the Club Méditerranée in Otranto," said an Italian official, "and last summer we had our best tourism season yet." At the time, my coauthor on this book called a chemical engineer with extensive experience in the field of hazardous transportation. "Tell me where the *Cavtat* sank," he said. "I'm not going within a mile of that wreck."

So far, an old story: Officials issued reassurances about the danger;

they dodged responsibility for the danger; they obfuscated the danger and even denied the danger. No one acted to defuse the danger. We wrote an article about the impasse, which in 1976 was published in the American magazine *Saturday Review* and was later reprinted in Europe by *Reader's Digest*.

That was when I met an extraordinary citizen who had stepped into the public forum.

As soon as the Cavtat had sunk in 1974, Judge Alberto Maritati, a young magistrate of the Otranto district, had immediately sought complete data about the toxicity of the *Cavtat*'s cargo. He had refused to accept the threat these poisons posed for his community. For two and a half years he had persisted in a letter-writing campaign to the appropriate ministries. He reacted to their response—deafening silence—by launching a one-man campaign to inform and arouse the public. With the publication of our 1976 article, tourism in the area fell. Maritati eventually gained access to the classified papers he needed to take legal action and sequestered the *Cavtat*, placing it and its deadly drums under his own court's jurisdiction. He went so far as to issue a court order to a privately owned recovery company to start a retrieval effort. In 1977, Maritati succeeded in forcing a major cleanup operation. My team filmed the exhaustive $12 million effort, which involved special equipment, specially trained industrial saturation divers, and ten months of labor. Today, decades later, children play on the beaches of Otranto. Their joy is justified and their safety is real.

There is, of course, always the alternative: maintaining the status quo. The prospect of a world with people persuaded that they have no right to judge risks and benefits for themselves—believing that they will advance if they swallow their objections and subordinate themselves to authorities—brings to mind the fate of an Andean tribe.

In 1968, we loaded our diving equipment onto a mountain train that rattled 12,500 feet up into the Andes, where we planned to explore Lake Titicaca and its Inca ruins. Out on the water's vast expanse we came upon a small group of natives, the Uru, remnants of an ancient people. The tribe had dwindled to only about two hundred Indians, huddled together in a way of life I'd seen nowhere else. The Uru built their villages on floating mats they fashioned from layers of totora, the papyrus reeds that grow in the lake's marshy shallows. While the Uru lived *on* the water, they also lived in fear of the water, believing that anyone who slipped in would die. They led their lives as their elders ordained, confined to their papyrus canoes and their papyrus islands, repairing their mats by laying fresh reeds on top as bottom layers rotted, raising cormorant birds like chickens for food, tending cooking fires on the dry totora reeds. When stray sparks jumped from their fires, the reeds of their papyrus islands ignited, and the Uru took to their canoes. There they waited until the fires burned out. Then they laid down fresh reeds to make another mat, another island, another village. Generation after generation dared the flames. No one dared the dictates of the elders. This trial by fire had never led the Uru to triumph.

It would lead, more likely, to extinction.

CHAPTER FOUR

IRREPLACEABLE WATER, IRREPLACEABLE AIR

B Y the time I was asked to serve as director of Monaco's Oceano-graphic Museum, majestically rising from its bluff two hundred feet over the Mediterranean, I felt I knew that sea almost as though it were a part of me. I had tested my Aqualung in its translucent waters, among its abundant life. I had sailed *Calypso* through its phyto-plankton meadows, watching from our underwater chamber as sunlight shone on these profusions of microscopic plants, base of the food chain, sustenance of the sea. When in 1957 I accepted the museum directorship, succeeding Commandant Jules Rouch, I could savor the Mediterranean from above as I had from below; my office windows commanded a view all the way to Corsica, its silhouette a pale silver in winter dawns, golden-red when bathed in the colors of rising summer suns. In those early days, I relished my morning routine: arriving at the museum well before it opened, walking down to the shore, and swimming along the Azure Coast, in the waters that the ancients called *mare nostrum*, our ocean.

Sixteen years passed. My predecessor, Commandant Rouch, died. The eminent explorer, a participant in the second Antarctic expedition of Jean-Baptiste Charcot on the famous *Pourquoi Pas?*, had requested burial at sea. Our forlorn group of mourners dutifully filed onto a boat docked below the museum and headed toward the proper depths for submersion. We traveled—and traveled—in mounting discomfort. Our boat was cutting through contaminated waters so foul that our senses re-belled at the thought of dropping a coffin. Not until we had just arrived at the prescribed location could our anxieties at least partially subside.

Only there—three and one half miles out from the shore where I once swam—did we finally find water fit for a corpse.

Decades have passed since Rouch's funeral, decades of journalists' questions to me and intermittent media alarms that, like trumpet reports, blast for an instant then fade from the air. Still businesses protest that pollution regulations disrupt the economy, even though international accounts prove that polluters loot the public coffers by running off with gross receipts and leaving costs to communities that must pay with massive cleanups, to farmers who must pay with ruined crops, to the people, who pay with their health. Still industries liberate unknown substances over Earth, testing to identify dangers only after they have exposed us, rather than testing to ensure our safety before. Towns continue to solve their problems by passing their poisons on to others, downwind, downriver, down the coast; nations continue to compute pollution in subtotals—asserting the "acceptability" of each of their individual infractions as they add relentlessly to the global aggregate. I have said that I am sick of questions about pollution. The truth is, I am sick at heart. The questions I hear still focus only on the relative disadvantages of an earth that is degraded—instead of on the unparalleled miracle of one that is not.

Why should we protect the environment? The more puzzling question concerns where our species got the idea that it is in the position to ask. Mere infants of the universe, with no feel for infinity, no sense of place in time and space, we human beings have yet to comprehend the enormity of what we are doing: In a geological second, we are unraveling complexities it took eternity to create.

What landmarks can we use as measuring sticks for the scale of time? The first written Greek histories of 2,500 years ago? The first time that humans succeeded in planting and harvesting food, 8,000 years ago? The Neanderthals of 85,000 years ago? These time spans are derisory in the majestic sweep of eons that led to life. Whatever an individual believes—that the Beginning was orchestrated by God or unleashed by blind forces—the exact chain of cosmic events that defined our origin remains beyond the reach of human dating and discovery. Our descriptions of that moment are all based on scientific supposition, inspiration, and dream.

While we record our written history in millennia, the history of life is writ in billennia, beginning perhaps fifteen billion years ago when all the universe consisted of nothing more than hydrogen. Its atoms collided and the alchemy began, with hydrogen producing helium; nuclear forces igniting the stars; helium burning to carbon and other elements; stars exploding, spewing their riches into other astral bodies, whose flames forged other elements, until one star became our sun, and the rocky grains around it fused to form our planets. Our solar system finally appeared—after perhaps ten billion years of cosmic cataclysms.

Inside the naked clot of matter that was to become the earth, rock began to melt; molten masses erupted through volcanoes, which belched out the stellar gases that had been trapped within. In this thick, primordial atmosphere, the gases condensed, the deluge began. Water drenched the planet, leached minerals from its crust and swept them to basins. There, in the warm proto-oceans, molecule interacted with molecule until they formed an unlikely molecular chain: one that could draw vital components from the mother sea to reproduce itself. Made of the stuff of stars, conceived by atmosphere and oceans, life appeared some one billion years after Earth itself had taken shape.

For perhaps two billion years more, deep within the sea, bacteria and simple cells of algae reigned as the only life on Earth, protected from the ultraviolet rays that blasted through oxygen-devoid skies. Slowly photosynthesis generated oxygen; slowly cells developed the machinery to protect themselves against this combustive gas and the genetic material to harness those combustive powers and make them their own. Just 600 million years ago, simple creatures at last produced complex progeny: corals and starfish with external skeletons, fish with spines inside, amphibia and reptiles. Only about 385 million years ago, vertebrates moved to land. Small mammals scampered among the dinosaurs more than 200 million years ago. Flowering plants blossomed only about 130 million years ago. Just 4 million years ago, on a planet that has existed a thousand times that long, an advanced hominid appeared. Only about 1.5 million years ago, he stood. In Earth's latest instant, just 100,000 years ago, the human brain attained its present size.

One frequently mentioned argument for protecting Earth refers to the fact that ours is the only known planet on which this sacred sequence took place, that Earth alone appears to be able to support the life we know. But the reason for safeguarding our planet transcends the simple reality that we're here—that we inhabit Earth. We are *part* of Earth—product of all that came before, kindred to all that now exists. Our bodies testify to the ties that bind us to the planet. Our spinal cord is but an elaboration of the simple nerve tube of a three-inch, eel-like sea creature, the amphioxus, believed to be the modern incarnation of the first vertebrate. Each of the twenty-eight bones of the human skull is a reflection of a bony part in the head of an early fish. Deep in flesh and bone we carry traces of even more elemental origins, bequests of primeval atmosphere and ocean—the iron in our blood, the carbonate of lime in our bones, the miniature sea inside each cell, less salty than the sea today but reminiscent of an era when some mutating fish turned to reptile and crawled to virgin land. We living beings—mammals, amphibia, reptiles, plants—we are air and water incarnate.

While the composition of our bodies suggests that atmosphere and ocean gave birth to life, each day the planet demonstrates the ongoing miracle, as atmosphere and ocean sustain life. The power of these two resources is all the more awesome given that, however "infinite" our supplies of air and water appear, only scant supplies of both somehow support our multitudes. Endless skies, bottomless seas, raging rivers, and violent rainstorms give us a false perspective. In fact, were all the waters on Earth combined—fresh and salt, liquid, vapor, ice—their volume would total just $\frac{1}{780}$ the volume of Earth. I once speculated that if our planet were the size of an egg, all its water would form no more than a single drop, barely enough to moisten the shell.

While I quickly calculated the quantity of water on Earth, I spent many years sailing the oceans before I recognized that only a small fraction of that small quantity of water supports life. Even as we began a study of the Mediterranean in spring 1955, I still found it easy to believe in the ocean's mythical bounty; we took nearly four hundred samples and found waters thick with the phytoplankton that feeds the sea. But after only one month of photosynthesis, even before the summer began, the microscopic plants had already exhausted the nutrients on the high sea; the watery meadows had almost disappeared. The Mediterranean, like a meager soil, could not sustain its crop for long. As I logged more miles in *Calypso*'s underwater observation chamber, I realized that the high seas are stretches of desert—living desert, to be sure, like land deserts, but with only a thin, hardly visible fog of plankton.

From time to time, the sea blossoms with life; there are luxuriant oases in the deserts of the sea, but those oases are rare. Life abounds in just three ocean areas: the top 160 to 320 feet, where sunlight meets nutrients to provide plant food for fish; the bottom, enriched by organic detritus that settles and decomposes; and especially, the continental shelves. Shallow ribbons of fertility, the continental shelves provide the ocean with its nurseries; here fish lay their eggs. The waters in these coastal areas constitute just one half of 1 percent of total ocean space, yet they support some 90 percent of all marine species. The continental shelves give the sea its literally narrow margin for life.

The relative scarcity of air on which all life must depend is even more unsettling. A spaceship rockets up only 124 miles to enter an absolute void. How much air is there in that 124 miles? Every schoolchild knows that with altitude, air becomes thin. If somehow we could compress the entire atmosphere to a consistent density of breathable sea-level air, it would be reduced to a height of only 4 miles. If we could further compress the atmosphere to the density of the oceans, we would see that the planet offers three hundred times less air than water.

Like the biblical two fish and five loaves that fed the multitude, these limited supplies of air and water nurture life. How? Of all the waters I have visited, the Caribbean Sea most vividly answers that question. Partly enclosed by a surrounding chain of islands, these generally poverty-stricken waters overall support relatively little life. But when we sailed into the Caribbean's eastern waters, off Venezuela, we entered Xanadu: There we watched resident baleen whales and manta rays thrive on an abundance of plankton, large fish prey on swarming sardines, sharks and dolphins roam the coral islands of Orquila and Los Roques.

Knowing that the Atlantic seeps into the Caribbean and fertilizes this area, we enthusiastically began an exhaustive scientific survey of the waters, testing as deep as six thousand feet, starting from the inflow's entry through the "door of the Caribbean," between the islands of Trinidad and Grenada, and following the flow all the way to Lake Maracaibo. East of Tobago, we were stunned by what we learned. The salinity and temperature of the water we drew unequivocally identified it as an Antarctic current! This very cold, heavy water sinks at the polar continent, then travels along the ocean floor—perhaps for a decade, perhaps for a century—gathering the Atlantic's bounty of fertile organic bottom detritus until an upwelling pushes it back to the surface. The icy polar water then nourishes the famished tropical sea.

Antarctic waters swirling over the equator: The phenomenon provides the solution to one of nature's riddles. Earth's supplies of air and water are minute, but they are governed by extraordinary forces—planetary currents and upwellings and winds that keep atmosphere and ocean in constant motion—reacting one with another, maintaining Earth's temperature and the sea's alkalinity and oxygenation, endlessly circulating as though they were a pulsing bloodstream and Earth itself a living organism. Finite and fragile, minuscule but majestic: air and water, the fluids of life.

These we pollute.

Into Earth's equivalent of a mere four miles of breathable air—the essential resource most quickly polluted—we spew industrial, auto, and airplane exhaust, pesticides, sulfides, heavy metals. Into Earth's drop of water we heave tons of sewage steeping in acid, volatile and semivolatile solvents, fuels, paints, partially combusted oils, potentially toxic halogenated hydrocarbons. Millions of sailors and yachtsmen on private boats and supertankers wait for night to cloak their own contributions, emptying their bilges of burned oil and tainted water that will plague the sea the next day as a slick. Nations dump the nuclear waste, the nerve-gas weapons—all the fiendish refuse they fear too dangerous to be disposed of on land.

Yet the tons of poisons dumped and spilled into the sea account for only 20 percent of marine pollution; land-based sources indirectly add four times more pollution than is disposed directly. Two thirds of acid rain falls on the oceans. The dominant source of lead and PCB ocean pollution is air pollution. Fully 25 percent of all the billions of pounds of DDT produced since 1943 is by now lapping in ocean waters. The sea is the global sewer, the ultimate receptacle, eventually glutted with all the pollution generated on Earth. Each time our population doubles, by some estimates, pollution multiplies six times—all of it into the fluids of life, all of it ending up in the sea. Out of sight, out of mind.

Few collective delusions are more insidious than this Pilatus syndrome: Dump it and wash your hands. Even today authorities continue to justify and even encourage ocean dumping by citing the sea's "dilution factor"; yet the truth remains that pollutants concentrate in the sea's most vital area—the shallows of the continental shelves, the cradle of marine life. Most of the world's ocean dumping has been routinely executed in waters no deeper than the North Sea, so shallow that were many of New York's skyscrapers posed on the bottom, they would rise above its waves. Winter runoff adds as much solid toxic waste to the continental shelves as does sewage disposal. Estuaries—ocean waters at each river's mouth, beneficiaries of the nutrient wealth of land—attract swarms of fish. Now those swarms feed on plants contaminated by the toxins that river currents deliver from all the nations along their course. While national spokesmen have been reporting reductions in single pollutants over single cities, the appalling addenda to reports as distinguished as those published by the Organisation for Economic Cooperation and Development have concluded that estuaries are "acutely damaged." After the tainted river plumes defile the estuaries, they next bend along the coastlines, continuing to discharge their toxic loads just off our seashores.

We not only foul the estuaries and continental shelves, cradles of life in the sea; we next pervert life's circulatory system. Logic dictates that powers capable of sweeping the Antarctic's fertility to the Caribbean likewise carry toxins to the remote corners of the planet. Again and again anecdotal evidence has turned this logic to reality: A particular type of poisonous pesticide sprayed on East African crops was detected only a few months later in the Bay of Bengal—four thousand miles away. With no nearby sources of atmospheric lead pollution, Greenland's polar caps collected lead at levels that rocketed 300 percent annually from 1940 to the late 1960s. Asbestos particles have lodged in Greenland's ice, after wafting in the atmosphere for months. Nature has given us tides that nourish the earth with nutrients; we have turned them to global tides of toxins.

In every area in which water is contaminated, so too are its creatures; after washing through the global sea and atmosphere, pollution percolates up the ladder of life. In the open water of the Antarctic, we came upon adult penguins, once slaughtered for their oil, now infected by chemicals from industrial cities on the far side of the globe. Overcome by DDT, they could no longer form solid eggs; their embryonic offspring died in the shell—a problem that has plagued the brown pelicans of Anacap Island, off California's coast; peregrine falcons; bald eagles; and other predatory birds. Pesticides concentrate in the tissues of Baltic herring and Pacific whales. Dead beluga whales have washed up on the shores of Canada's St. Lawrence River with more than thirty hazardous chemical pollutants detected in their tissues. PCBs, among the most powerful toxins known, have lodged in the blubber of seals in the Gulf of St. Lawrence, in the Baltic, and off Britain's coast. PCBs, in fact, have been detected in all fish species tested, an unnerving index to pollution's widespread dissemination.

The sea returns to us that which we give it. If we protect it, it protects us; if we abuse it, it returns the abuse. It would be foolish to believe that pollution festers in the tissues of all species except the one that generates it. Pesticides have been detected in virtually every human being on whom a test has been conducted. Most of these stable poisons cannot be retrieved, not from the sea, not from its creatures—which return the toxins to the oceans when they die—not from human tissue, not from mother's milk. Out of sight, out of mind—but ever present.

Fifteen billion years for star dust to produce a living planet. Twelve years for the Mediterranean below my window to turn from water that invigorated a swimmer to water too filthy for a corpse. For the human species, time has moved both too slowly and too fast. Throughout the past million years, human beings could survive only by continually intensifying their violence; individuals held their grip on life by warring against nature's threats of death—carnivores, cold, lightning, fire, and flood. Only recently in the course of their existence, by sheer cunning, human beings have managed to match nature's power and even overwhelm it. Yet victory has come too suddenly. Habit compels us to intensify the struggle; reason has yet to convince us that we face no more enemies, other than ourselves. After a million years of accelerating human force, we're now unable to make the necessary hairpin turn.

Ironically, despite the intellect that sets us apart from Earth's living animals, we may fare no differently from the animals that lost their contest with extinction. Paleontologists contend that no species naturally develops a harmful structure, that instead, once-successful species disappear when they fail to adapt after circumstances change. The human being, capable of invention but incapable of restraint, could find intellect not only his prize attribute but also his fatal flaw. With his unadaptive attitude, the human being could become the Cenozoic era's dinosaur.

Earth's past has much to teach; its future gives us something to contemplate. Astrophysicists calculate that we have appeared on Earth at just about its halfway point, that the planet will exist another four or five billion years. Whether our own newborn species, just one million years old, lives only as long as some fossil invertebrates did—ten million years—or whether human beings five billion years hence see the fiery spectacle at the sun's end that went unwitnessed at its beginning, who can say? Modern knowledge can progress no further than the wisdom transcribed on a Dead Sea Scroll found fading in the Judean desert: "None there be, can rehearse the whole tale." The time that lies ahead—with us or without us—may tempt scientific calculation, but it defies human imagination.

If indeed existence, for our Earth, unfolds as far into the future as it has from the past; if five eons ago an astral cinder began a metamorphosis to wet blue globe; if in the intervening eons the oceans brought forth living beings, animalized water—then what wonders await in the five eons to come, what unexpected geological cataclysms, what continental shifts, what mountains moved, what new genetic combinations, what new creatures with what new capacities, will be produced by Earth's molten core in the five billion years ahead?

With faces turned toward this future, humans of just one generation or two commit irreversible environmental acts. They rationalize that they must do perpetual damage to Earth to keep their corporations financially robust, to enhance their national trade balance, to promote their political popularity. After all the centuries of civilization, the most resonant response to this attitude remains the remark made by the Greek philosopher of nature Anaxagoras. When legislators criticized him for distancing himself from politics, for showing more interest in his scientific studies than in his country, the sage pointed to the stars. "There," he said, "is my country."

It little matters whether the old man really sensed our birthplace out in the universe. The importance of his comment lies at the point where philosophy and science meet. If our country is the cosmos, then Earth is its Emerald City, the breathtaking speck where astral embers turned to water, air, and life. Never again can life appear anew on this planet; the oxygen that sustains life today would be so corrosive to cells without the protective mechanisms that they evolved in eons gone by that re-creation in the presence of the gas would be impossible. Probably never again can human life appear anew elsewhere in space, even given the odds of infinity and eternity. Liquid water is essential to life. Elements most often exist in gaseous or solid state. Water is the rarest element of all, remaining liquid in only a very narrow range of temperatures. The exact distance of our planet from its sun, where water neither boils nor freezes, the intricate chain of molecular combinations and recombinations that turned water to cell, cell to human—it is miraculous that these events occurred, inconceivable that they could reoccur.

Why protect Earth's biosphere? Each day that we ask the question, rather than act on it, some requirement for life, some necessity for the quality of life, some living being is lost or threatened. The future we are creating may be one of miracle or tragedy: None there be, can rehearse the whole tale. The question is not, why must we protect the biosphere; the question is when. The answer is now.

CHAPTER FIVE

THE HOLY SCRIPTURES AND THE ENVIRONMENT

O NE Sunday in the early sixties, a long series of violent storms that had been raging against Monaco finally abated. I had recently designed a floating ocean-research station, and we'd managed to moor a prototype in the deep Mediterranean halfway between Monaco and Corsica. For the previous month, the storms had cut off all access to it. The scientists on board the floating research station, I knew, were running precariously short of fresh water and food. Certain that the calm would last the afternoon, I hastily loaded provisions onto a hydrofoil.

Because the boat could hold seventy-five people, I invited my staff at the Oceanographic Museum, together with their families, to join me on my trip to the station. Christened Mysterious Island after the novel in which Jules Verne concocted fanciful ocean inventions, the "island" consisted of a huge, upright, buoyant hollow cylinder, one end rising above the surface and the other penetrating two hundred feet under the sea. Interchanging teams of six scientists lived aboard in a small house perched on the top of the cylinder, above the surface. By means of an interior elevator, they descended down through the cylinder into the depths, where all night and day they measured marine life, water evaporation, currents. Because Mysterious Island had become somewhat renowned as the first manned research buoy anchored in deep-sea waters, my entire staff eagerly accepted my offer to show them the laboratory.

About an hour after we had left the shores of Monaco, I saw a shape emerge on the horizon: the island. I announced our approach, and my friends excitedly rushed onto the deck for even this first, distant vision. But one boy, the thirteen-year-old son of a museum administrator, lingered behind. He was reading a comic book. I repeated: "You can see Mysterious Island from here!" The boy looked up in distracted acknowledgment of my presence. He didn't budge. Soon he was reimmersed in his comics.

The exclamations of the other passengers wafted in to us. Still the youngster's nose was buried in his book. I stepped over to see what held him so spellbound.

He was poring over a cartoon story about my Mysterious Island.

The boy favored reading over reality. Adults might have characterized him in any number of negative ways—as uninquisitive, uninvolved, apathetic about the world around him and his place in it. I've often wondered: Are many adults much different when they read the scriptures of their respective faiths?

Ninety percent of French people are Roman Catholics, as are ninety-nine percent of Italians. Germans proclaim "God with Us" as their national motto, just as "In God We Trust" is the motto chosen by Americans, half of whom say they attend church. Fifteen million people ascribe to Judaism; 200 million, to Buddhism; 400 million, to Hinduism; 500 million, to Islam. How many of these people rise to their feet or fall to their knees in cathedrals, temples, synagogues, mosques, reciting the word of their God by rote, all the while ignoring the living word of God just outside the window? How many read scriptures that praise their God's creation but acquiesce when damage is done to it? Daily newspapers report on politicians, presidents, ayatollahs who righteously and regularly proclaim that they lead their nations in accordance with the word of their God; we hear of martyrs who have died because they have refused to repudiate their beliefs, of revolutions, civil wars, holy wars—all waged by people who are willing to fight for the right to believe what they choose. They choose to believe in a God who has issued divine commands; how many honor His divine commands to safeguard the environment? How many instead behave as latter-day Peters, vociferously attesting to their belief in God but denying Him when the opportunity arises to protect the environment as holy writings mandate?

The endurance of the scriptures' message makes their call to safeguard natural resources carry significance even for those who do not believe in God. The prophets began to preach what became the essence of various scriptures about the time Ramses II built the temple of Abu Simbel. Crude hieroglyphic birds and fish notched into the pharaohs' temples at that time have long since been forgotten. Even the insights of Aristotle, Archimedes, and Sophocles illuminated civilization like flickering candles, shedding momentary light before fading from central relevance. All the while the ideals of the scriptures have remained compelling through a sweep of millennia that buried most other masterpieces. Those ideals clearly include environmental protection. The scriptures transmitted by God according to Jews, Christians, Muslims, Hindus, Buddhists, Taoists, and Confucianists—*half the population of the earth*—are clear: The glory of nature provides evidence that God exists; those who show no respect for nature show no respect for God. "They are without excuse because when they knew God, they glorified Him not as God," wrote Saint Paul. "Professing themselves to be wise, they became fools . . . For the invisible things of Him from the creation of the world are clearly seen."

Faith after faith exhorts its followers to open their eyes to nature as a reflection of God's grandeur. Solomon lamented in the Apocrypha, "Men were unable from the good things that are seen to know Him who exists." In the Koran, Allah declares to the Muslims, "And as to the earth, we have spread it out, and have thrown the mountains upon it, and have caused an upgrowth in it of all beauteous kinds of plants, for insight and admonition to every servant who loveth to turn to God." The Muslim prophet Muhammad also repeated the principle: "Verily, in the creation of the Heavens and of the Earth, and in the succession of the night and of the day . . . are signs for men of understanding heart; . . . 'Our Lord!' they say, 'thou hast not created this in vain. No. Glory be to Thee! Keep us, then, from the torment of the fire.' "

Hindu scriptures go further: Not only is the existence of God evident in the wonders of nature; God *is* nature. In the epic *Bhagavad Gita*, the Hindus' supreme devotional book, the Blessed One proclaims, "On me all this universe is strung, like heaps of pearls on a string. I am taste in water . . . I am light in the moon and the sun . . . The goodly odor in earth . . . Of bodies of water I am the Ocean; I am Varuna [god] of water creatures . . . I am the dolphin of water monsters; of rivers I am the Ganges."

Throughout time and across the bounds of countries and cultures, religions have also held the environment sacred as the source and substance from which God created His greatest gift—life itself. The Greeks named their goddess of nature by combining the words for "earth" and "mother": Demeter. Plato wrote that the earth should not be compared to woman but that woman should more rightly be likened to the earth; long before human beings existed, the earth, when fertilized, bore fruit and nursed her fields on river waters. Even today, Peruvian Indians celebrate births by laying their infants on the ground so that Pachamama, the earth mother, can cradle them. American Indians also revere the earth as maternal. One legendary Indian sage refused even to plow his fields: "Shall I take a knife and tear my mother's bosom? Then when I die she will not take me to her bosom to rest. Shall I dig under her skin for bones? Then when I die, I cannot enter her body to be born again."

The scriptures of organized religions maintain the same tenets. "Dust thou art and to dust shalt thou return," states the Bible, with the original using a Hebrew word for dust that translates more precisely as "pieces of earth." Muhammad, whose words Muslims regard as higher truth than science, declared: "And God caused you to spring forth from the earth like a plant. Hereafter will He turn you back into it again and will bring you forth anew. He well knew you when he produced you out of the earth, and when ye were embryos in your mother's womb."

Scientists cite evidence that the first living cell, composed of elements drawn from the smoldering infant planet, gestated in the rich broth of the early ocean—that life was in fact born of earth and nursed by sea. Yet thousands of years ago, on the basis not of any scientific finding but of inspiration alone, the author of Genesis wrote, "And God said . . . Let the waters bring forth abundantly the moving creatures that have life, and the fowl that may fly above the earth in the open firmament of heaven. And God created great whales, and every living creature that moveth, which the waters brought forth abundantly, after their kind, and every winged fowl after his kind."

Muhammad too foresaw the scientific principle of today, stating it as a holy principle of all time. He urged his followers to revere water as the womb of life. "And it is He who hath created man of water, and established between them ties of kindred and affinity. Do not the infidels see that . . . by means of water we give life to everything? Will they not believe?"

Holy writings furthermore emphasize that after using water to conceive life, God uses water to sustain life. "He it is who sendeth the winds as the forerunner of his mercy rain; and pure water send we down from Heaven, that we may revive by it a dead land," wrote Muhammad. "And we give it for drink to our creation, beasts and men in numbers; and we distribute it among them on all sides, that they may reflect: but most men refuse to be aught but thankless."

In one of the New Testament's most inspiring miracles, Christ called on the bounty of the sea to provide for his disciples:

> Launch out into the deep, and let down your nets for a draught. And Simon answering said unto Him, Master, we have toiled all the night, and have taken nothing: Nevertheless at thy word I will let down the net. And when they had done this, they enclosed a great multitude of fishes: And their net brake. And they beckoned unto their partners, which were in the other ship, that they should come and help them . . . And when Peter saw it, he fell down at Jesus' knees . . . for he was astonished.

During my team's expedition to Tierra del Fuego, the mountainous archipelago on the extreme southern tip of South America—in the harsh reality of life at its most brutal—I learned for myself just how instinctively human beings sense the sanctity of water. The sole survivor of the lost tribe of the Yahgans told me about the tales and rituals of her vanished people, recounting stories that showed how profoundly her tribe had revered water as sustenance of both body and soul.

I had first heard of the Yahgan tribe of Tierra del Fuego years before, from a priest named Abbé Lempereur. One afternoon he came to visit me in my apartment in Paris. Ignoring the city noises that welled up from the busy streets below, I was transported by the priest's tale to a distant place and time, an era when the hearty tribal Yahgans ruled the rugged terrain of Tierra del Fuego.

Tragically, when Charles Darwin described the Yahgans in the 1830s in notebooks he kept on his historic voyage aboard the HMS *Beagle*, the Indians of Tierra del Fuego lost their anonymity, and with it, their quality of life and even their hold on life. Darwin returned to his home shores with animated accounts about grotesque human animals. "I never saw more miserable creatures," he wrote, "stunted in their growth, their hideous faces bedaubed with white paint & quite naked . . . Their red skins filthy & greasy, their hair entangled, their voices discordant, their gesticulation violent & without any dignity. Viewing such men, one can hardly make oneself believe that they are fellow creatures placed in the same world." Darwin compared the Yahgan men and women to monkeys.

More than one hundred years later, Abbé Lempereur was less concerned about the supposed barbarity of the Yahgans than about the barbarity of the Westerners who had come supposedly to civilize them. The emissaries of the cultured world shot the natives like game birds. They infected the tribe with their insidious imports of syphilis and alcoholism. Of the thousands of Yahgans who had greeted the first explorers, Abbé Lempereur told me, only twenty-four people survived. The explorers of the 1800s, who had planned to civilize the savages, had savagely nearly eliminated them.

Abbé Lempereur had devoted his life to staving back the complete extinction of this endangered people. But by the time I was able to organize an expedition halfway around the globe to their windswept cluster of islands, all that was left of the once robust Yahgan tribe was one ailing, wrinkled, eighty-two-year-old woman. The history of a whole people rested only on her fading memory.

Even a century after Darwin's ugly description of hardship, life on her island had been difficult. The tribe had rolled newborns in the snow, a triage method of selecting the best infants: the ones who survived.

While the men tended the children, the women dived for fish; the sea provided most of the people's nourishment.

Waters also nourished the spirit of the Yahgans. The tribe believed that the earth was a small and fragile boat adrift on the surface of a vast ocean universe. The whale ruled this liquid cosmos. Yahgans celebrated every significant milestone of their lives by paying homage to Leviathan. The tribe would even postpone marriages and initiation rites in the hope that a whale would strand itself onshore. When one of the behemoths finally would beach itself, the people believed that the animal was not really helpless on the sands but rather that their shaman had persuaded it only to feign helplessness in order to help them. Tribesmen would cut the great mammal into portions and use the meat in religious ceremonies, believing that the whale would miraculously re-create itself and swim back to sea after it had nourished each and every one of its human disciples.

Throughout scriptures sacred to the world's faithful, water is revered not just for its bounty but for its purity as well. Water washes away the ills that ravage the body and the sins that stain the soul. The baths that the people of Jerusalem called Bethesda healed the halt, the lame, the blind. "An angel went down at a certain season into the pool," wrote the disciple John. "Whosoever first stepped in was made whole of whatsoever disease he had."

Ganga Ma, Mother Ganges, the redeemer goddess of the river that Hindus hold sacred, restores the withered spirit as Bethesda restored the withered limb. She wanders in winter through Shiva's hair—the frozen Himalayas—until spring, when she rushes through the snow peaks and flows to the sea. Hindus say that once, on her springtime journey, she paused to wash away the sins of sixty thousand lost souls. If only a few drops of her waters are placed on the tongue of a dying man, they will redeem his soul. "O waters . . . ," Hindus pray, "let us resort to you full for the removal of evil, whereby you gratify us."

Christians celebrate their sacrament of baptism in reverence for Christ's request that his soul be renewed by the waters of the River Jordan. He waded into its waters, "and the Spirit of God descended like a dove . . . ," the Bible recounts, "and lo, a voice from heaven, saying

This is my beloved Son, in whom I am well pleased." Can this account have no relevance for those who dump their chemical wastes in streams and rivers and lakes, polluting water, the very liquid with which Christian believers must be washed if they wish to enter the kingdom of heaven?

The unrepentant, according to holy scriptures, have discovered that water, wellspring of God's mercy, also serves as instrument of His wrath. "God saw that the wickedness of man was great in the earth . . . And it repented the Lord that he had made man on the earth, and it grieved him at his heart." Came the Deluge.

Throughout the ancient world, sages told astonishingly similar accounts of a great flood, an unyielding wash of water over a hopelessly renegade population. Muhammad relates the words of Allah: "I utterly abhor your doings . . . And we rained a rain upon them, and fatal was the rain to those whom we had warned."

The significance of the flood lies not just in how many lives were lost but also in which lives were saved. "And the Lord said unto Noah . . . of every clean beast thou shalt take to thee . . . the male and his female . . . ; to keep seed alive upon the face of the earth." Humanity's foremost obligation in the flood, as ordained by God, was to preserve biodiversity, to ensure that not one single species vanished from existence. The importance of the command is underscored in the Bible's emphasis that God entrusted the task to the only righteous man left on Earth.

What of the modern contention that some creatures are of no "use," that they in fact ought to be sacrificed on the altar of development and profit? Christ addressed the contention two millennia ago: "Behold the fowls of the air," He said. "They sow not, neither do they reap, nor gather unto barns; yet your heavenly Father feedeth them." He asked other followers, "Are not two sparrows sold for a farthing? And one of them shall not fall on the ground without your Father."

Buddha too issued an unequivocal command to safeguard all species:

Whatever living being there be, large, medium, or small, developed or developing, may they all be at ease. Let no one humiliate or despise

another, anywhere whatsoever . . . Develop boundless regard to all
beings, even as a mother would cherish her only child.

Scientists report that plant and animal species are now disappearing,
at the hand of humankind, far faster than nature can replace them with
new creatures. Biologists should find allies in their quest to prevent
more extinctions in the billions of people who muster their strength
in churches and mosques and synagogues and temples, a file of faithful
that stretches across the globe.

Some scholars, it must be said, disagree. Academicians have argued
that the Bible, rather than urging believers to cherish the environment,
actually encourages them to plunder it. The scholars who argue that
the Bible is anti-environmental support their contention by citing two
verses.

Replenish the earth, and subdue it: and have dominion over the fish of
the sea, and over the fowl of the air, and over every living thing that
moveth upon the earth.

[Thou gavest man] dominion over the works of thy hands; thou hast
put all things under his feet: All sheep and oxen, yea, and the beasts of
the field: The fowl of the air, and the fish of the sea, and whatsoever
passeth through the paths of the sea.

Indisputably, these passages assign jurisdiction over nature to human beings; the verses can even be interpreted as instructing humankind
to reap the benefits of Earth's resources. Who could rationally protest?
We are blessed with the opportunity to live off the interest gained from
nature's capital.

It is the exhaustion of nature's capital, the irreparable destruction of
her limited natural resources, that the scriptures condemn. Those who in-
terpret the Bible as commanding the bankrupting of nature's accounts
have offered up a *contre sens*—a translation that says the opposite of what
was intended. The scholars base their argument on the supposed biblical
directive to "subdue" the earth and to take "dominion" over animals.

Experts in exegesis, however, point out that in the original Hebrew, the intent of the author is more clear. The word *dominion* was used to translate the Hebrew word *rada*, which refers to the guardianship offered by a shepherd king. The prophets used the word *rada*, in fact, to describe the nobility and magnanimity of King Solomon's rule. A command to reign over nature in such a way is a command not to vanquish it but to protect it as a great sovereign would protect his people. As for the words *subdue . . . the earth*, the word *subdue* was used to translate the original Hebrew word *kibbes*, which refers only to soil and plants and is used specifically not for any kind of devastation or destruction but for pressing grapes for wine. The verse, moreover, commands human beings to subdue *and to replenish* the earth. Even more specific are the poetic words of Solomon himself in the Apocrypha:

> O God of my fathers, and Lord of mercy, who madest all things by
> thy word, and by thy wisdom formest man, that he should have
> dominion over creatures that were made by thee, and rule the world in
> holiness and righteousness, and execute judgment in uprightness of
> soul: Give me wisdom . . .

The verses bring to mind the benign monarchy of legend, in which the king safeguards his subjects in the belief that it is he who in fact serves them, that they themselves reign supreme. Human beings have been bestowed the royal privilege of protecting nature; nature rules. When God promised never again to destroy life with water, it was to His planet—not to humans—that he offered the splendid symbol of his pledge: "I do set my bow in the cloud, and it shall be a token of a covenant between Me and the earth." Another Old Testament sage beseeched, "When I look at thy heavens, the work of thy fingers, the moon and the stars which thou hast established; what is man that thou art mindful of him?"

An American Indian I once met recounted an experience that showed how firmly he grasped the principle that human beings are caretakers, not possessors, of nature. *Calypso* had sailed the rivers of Canada into the Great Lakes. One summer evening, walking along the shoreline, we met the Indian, from the Montagnais tribe. He invited us to share his dinner, salmon grilled over open flames. Sitting by the dancing flames of the fire, as dusk settled over his campsite, the Montagnais spoke of an American man who had recently purchased land along a nearby river and had subsequently prohibited the Indians from fishing there. The river, the interloper had said, now belonged to him. The idea that a human being could own a river had amused the Indian. "If the river is yours," he had responded to the newcomer, "take it with you to the United States."

In fact, this member of the Montagnais tribe had only been repeating a lesson that holy scriptures have taught through time: "The land shall not be sold forever," states the Bible, "for the land is mine; for ye are strangers and sojourners with me." Jehovah warned even the steadfastly faithful Job to guard against the conceit that the earth was made exclusively for human benefit, or that humans could possibly claim preeminence over creation:

> Who has cleft a channel for the torrents of rain, and a way for the thunder bolt, to bring rain on a land where no man is, on the desert in which there is no man; to satisfy the waste and desolate land, and to make the ground put forth grass? . . . Where were you when I laid the foundation of the earth? . . . Doth the hawk fly by thy wisdom and spread her wings toward the south? . . . Who shut up the sea with doors, when it brake forth, as if it has issued out of the womb . . . and said, Hitherto shalt thou come, but no farther: and there shall thy proud waves be stayed?

Clearly the God these scriptures describe would not command humans to pillage nature, to profane His handiwork. On the contrary, humans are made in the image of God, the Creator; plunder is blasphemy. When Jehovah bestowed on humans the privilege of having dominion over the earth—as when he placed Adam in the Garden of Eden to "till and keep it"—he ennobled us. He entrusted us with the guardianship of His creation.

The scriptures do not stop at symbolism, allegory, or even commandments in their message that human beings must protect the environment.

Scriptures state outright that the quality of human life on the earth depends on the way that humanity treats the earth. Buddhists teach that environmental degradation parallels moral degradation. When the human character is at its worst, Buddhist priests say, the environment is in turn damaged and human life expectancy decreases as a result. Conversely, Buddhists contend that a protected environment produces a healthy and even a moral society.

Confucianism and Taoism are subjects close to my own heart; for my admission to the French Naval Academy at Brest, I was given three hours to complete a dissertation on the two philosophies. Both, I wrote, teach that the happiness of human beings depends on their treatment of the environment. Taoism holds that humans should actually emulate nature. "Being in accord with Nature," Lao-tzu wrote, "he is in accord with Tao; being in accord with Tao, he is eternal; and his whole life is preserved from harm." At the sanctuary at Ise, Japan, the holiest of Shinto shrines, perfectly cultivated rice paddies edge up to the border of virgin forest, illustrating the respect that humans can accord the wild even as they gently coax productivity from adjacent soils.

Lao-tzu warned against ignoring these principles. "There are those who will conquer the world and make of it what they desire," he taught. "I see that they will not succeed. For the world is God's own Vessel. It cannot be made by human interference. He who [interferes with it] spoils it. He who holds it loses it."

Muhammad too forbid his followers to ravage nature. "Bear in mind the benefits of God," he wrote, "and lay not the earth waste with deeds of license."

Sacred writings brook no exceptions to their prohibition of environmental destruction, even in the case of war. "When thou shalt besiege a city," wrote an Old Testament scribe, "thou shalt not destroy the trees thereof . . . for the tree of the field is man's life."

The prophet Isaiah referred caustically to wartime deforestation. When the detested Nebuchadnezzar was vanquished, Isaiah taunted the defeated king for having cut down the cedars of Lebanon. "The whole earth is at rest, and is quiet; they break forth into singing, yea, the fir tree rejoices at thee, and the cedars of Lebanon, saying, Since thou art laid down, no feller is come up against us." Shintoism and Confucianism hold that a collapse of the harmony between human beings and nature actually provokes war—an ancient belief with ominous modern pertinence in a world in which human populations are exploding, in which wealthy nations squander scarce resources and the deprived and disenfranchised simmer in resentment.

And what of those who ignore the holy warnings? What of those who raze the rain forests, poison our water wells, fill the air with fumes so noxious that dead birds fall from our skies and clouds pour down acid rain? Wrote Muhammad: "Greater surely than the creation of man is the creation of the heavens and of the earth; but most men know it not. These are they who have lost their souls, for that to our signs they were unjust."

In the Koran, Allah inveighs: "Have we not made the earth a couch? And the mountains its tent-stakes? And we send down water in abundance from the rain-clouds, that we may bring forth by it corn and herbs, and gardens thick with trees. Lo! The day of Severance is fixed . . . for . . . they gave the lie to our signs." The Koran continues: "Hell is before him: and of tainted water shall he be made to drink: He shall sup it and scarce swallow it for loathing; and death shall assail him on every side, but he shall not die: and before him shall be seen a grievous torment."

Leaders who have led their people into permitting or even promoting the pollution and depletion of Earth's bounty—including those who nonetheless vaunt themselves as upright, moral, God-fearing men—have reason to fear God. "Behold," cries Isaiah, "the earth is utterly broken down, the earth is clean dissolved . . . and it shall fail, and not rise again. And it shall come to pass in that day that the Lord shall punish the kings of the earth upon the earth. And they shall be gathered together, as prisoners are gathered in the pit."

Nor will the wrath of God be limited to wayward leaders. The prophet Isaiah spoke to common citizens who "sin by silence," to quote Lincoln, "when they should protest."

"O sinful nation," Isaiah wrote, ". . . your country is desolate, your cities are burned with fire: your land, strangers devour it in your presence, and it is desolate . . . And the destruction of the transgressors and of the sinners shall be together, and they that forsake the Lord shall be consumed . . . And the strong shall be as tow, and the maker of it as a spark."

The last phrases of the verse: ". . . they shall both burn together, and none shall quench them."

CHAPTER SIX

SACCAGE

IN the late sixties I began to sense that something besides pollution was compromising life at sea. Coral reefs, for one, were shrinking. Pollution could explain the retraction of reefs that were washed by murky, poisoned waters near industrialized areas. But the great coral reefs in the Mozambique canal, lapped by crystal waters, were also shrinking, as were New Caledonia's corals in their glassy sea.

What suddenly could be overpowering corals that had sustained and even flourished under the onslaught of Earth's progressive ages? Every time we had ever followed a coral wall down into the depths, we had experienced a sense of traveling back in time. Primordial seas gave birth to the first corals six hundred million years ago, when simple, single cells first began to grow complex. Tiny coral polyps, distilling limestone from the sea's warm waters, formed external skeletons, then fused together, multiplying and mounting higher, half an inch a year. As the ages progressed, so too did they, unfurling into ever more splendid shapes, with the notches and caves and crevices they formed providing shelter for a profusion of other creatures, until corals and their exotic occupants became the most elaborate ecosystems of Earth.

And so it was, that sailing south of India among the Maldives, we maneuvered *Calypso* into a lagoon—a coral cup filled to its rim with translucent waters. Day after day we filmed evolution's resplendent display—corals opening into shapes like lacework fans and flowers, branching into staghorns and elkhorns, glowing under our lamps in brilliant hues of red and orange and innumerable pastels. Within this dense coral growth thrived myriads of the ocean's own jungle creatures: unicorn fish and rabbit fish, porcupine fish, clown fish, and damsel fish. Moray eels lurked behind coral bushes; rays, primitive cousins of the shark, glided through liquid skies with schools of barracuda and jacks. From April through August we spent our days astonished, below the surface, and our nights serene aboard *Calypso*, floating motionlessly on the lagoon's limpid waters, protected from the pounding sea by the paradoxical coral polyps—individually fragile but collectively strong enough to stave off waves that crashed against the reef with an average force of 500,000 horsepower.

Yet the reef could not stave back another force. One day we surfaced from our dive to see the Maldivians toiling in the sun. With neither wood nor stone on their sandy paradise, they built their houses out of coral. All year round they labored on the reef, hacking out blocks that they mortared together by grinding more coral to make cement. They raised their houses by demolishing their home—the very reef that protected them from the impact of ocean waves. Several islands, undermined by their own inhabitants, had actually vanished, swept away and swallowed by the unforgiving oceans. The islanders had, figuratively and literally, sawed off the coral branch on which they were sitting.

At the time, I saw the natives' mechanical destruction of their reef as a tragedy, but only an isolated one. Although my friends and I did not condone the Maldivians' behavior, we could understand it; we once had acted in the same way. The sea's riches seemed so abundant to us during our early years; taking them seemed so natural. We could reach out with our hands and grab a fish, and so we did. We could easily skewer fish with spears, crossbows, harpoons, and so we did. I even used to brag about the time I won a wager with a friend by catching a fish on a knitting needle.

And so I continued to deceive myself. If there were no fish to film in one area, I pushed off to another. To the next cape, to the next bay, around the next corner. By the seventies, we were turning corner after corner to no avail. The rocky shores of Brittany as well as the Canary Islands in the Atlantic had been progressively depleted. I returned to Assumption Island in the Indian Ocean, whose undersea gardens had in the early fifties given us the opportunity to forge a friendship with the grouper we called Ulysses, whose dances with us made him a celebrity in the film *Silent World*. When we had filmed the waters around Assumption Island those many years before, they had so burst with life that my divers described them as "the sea turned inside out." The area now was barren. I thought back to all the other films I had completed; I could no longer have remade a full dozen of them. The devastation had progressed too far.

About this time, people around the world were beginning to recognize and rally against the threat of pollution. Spurred by public awareness, some governments declared themselves ready to take action. This was the moment, I thought, to use *Calypso* to make a rigorous study of the Mediterranean. A sea almost completely enclosed by land—a kind of scale model of the global ocean—its ills often predicted those of worldwide waters. If we could detail the full extent of the Mediterranean's pollution, perhaps we could avert some global problems and even identify solutions. At the time, I was serving as secretary general of the International Commission for Scientific Exploration of the Mediterranean (CIESM), which agreed to sponsor us. The United Nations Environmental Program also joined the project and would eventually copublish the findings.

And so we set off on a five-month journey along the coasts of twelve Mediterranean countries; we stopped in more than 140 places to gather water, sediment, and plankton samples, which scientists at our Monaco Oceanographic Museum would test for toxins.

Measuring only pollution levels, however, seemed inadequate to me. I wished there were a way to measure levels of life, to grade the sea on some kind of "vitality index." While I knew there was no scientifically valid method of measuring life statistically, I thought that, as divers and as filmmakers—as people who had seen firsthand the changes that had occurred in the Mediterranean during the past two decades—we could assess life subjectively in a way that no one had before. We dived to 197 feet and used our diving saucer to descend to 1,150 feet. We sent the troika, our mobile underwater camera sled, bounding along the bottom, filming the conditions of the seafloor. We compared our new films of underwater areas with films we'd taken in the same places decades before. We spoke with fishermen in Spain, Algeria, Tunisia, Romania, Italy, Greece, and France. We talked as well with presidents and prime ministers, secretaries of fishing ministries and of the environment, industry leaders and owners of fishing fleets.

Everything we saw and heard led to one unequivocal conclusion: Life in the Mediterranean was diminishing. Groupers, corbs, dentrex had nearly disappeared; lobsters were smaller than those we'd found at the beginning of our career and inhabited deeper waters; "esquinades"—sea spiders—no longer came to shore in May to spawn. Even pods of dolphins, which had played so joyfully and effortlessly beside my Navy cruiser decades before, had diminished. Nations that sponsored our project offered to declare sixteen sites eligible as potential marine parks, portions of the littoral to be set aside so that their creatures could recover. All the sites, however, were already so devoid of life that such seventh-hour protection would have been meaningless.

I estimated that 30 to 40 percent of Mediterranean life had disappeared.

We steeled ourselves to receive the pollution measurements on the water samples we'd sent to the lab. The results were puzzling. Chemicals did indeed swirl around river outlets, coastlines, and urban shores. Yet effective pollution laws had significantly reduced some of the most offensive toxins, like pesticides. The bleak shores and forsaken waters that we'd seen could not be explained by pollution alone. Something else had to be at work.

With a heavy heart I admitted to myself that I had long sensed the answer. As a young diver, I had plucked lobsters from the depths with the naïveté of a child plucking flowers. I had regretted and corrected my mistake. But the mistaken complacency that I myself had once felt had lived on in the community, had multiplied and ramified, had lost its innocence and grown ugly. The Mediterranean was now under mechanical assault from all sides. Hundreds of millions of people inhabited coastal countries, with populations in the most fragile arid zones having doubled in the previous two decades. One third of the entire world's tourists swarmed to its beaches. To satisfy the demands of these swelling crowds, countries had turned against the sea. They had destroyed or al-tered significant portions of the shallow coastal waters, smothering life and mutilating habitats with dikes, landfills, artificial harbors, and the half-closed, stagnating waters of man-made beaches. Entrepreneurs also invaded the shores, crowding them with hotels, marinas, yacht ports. They had quadrupled the docking space they had provided for pleasure craft, devastating coastlines to make room for more boats. While zoning regulations had forbidden some construction abuses along beaches, skyscrapers had risen at water's edge, with the highest officials paid bonuses by the floor for turning their eyes. Industry, too, raced to buy up property at the shoreline, where it used coastal waters as cheap cooling ponds and refuse dumps.

All around the planet I've since seen the same behavior. Wetlands, dried out. Rivers, diverted. Mining and trawling, shooting and spearing, dynamiting, drilling, and dredging. We all are Maldivians, hacking and cutting and carving away at Island Earth. Nature's irreplaceables are being plundered—ransacked, pillaged, looted—by the marauders of our mod-ern age. Perhaps even more than pollution, mechanical destruction—I call it *saccage*—is severely damaging the sea's environment. Pollution is often the result of negligence or ignorance. But saccage is a more deliberate aggression.

Those who oppose all development—the extremist environmental-ists who respond to every plan with "No!"—compromise the future in their own way. Yet I long for a place for those who say, "No, but . . . ," those who propose rational ways to benefit from nature's dividends and refuse to squander its capital.

Even more counterproductive to society than those who reject everything are those who stop at nothing. They urge us to accept the one supreme fallacy that fuels saccage—that we must permit the plunder of the environment "for progress"—as if destruction were advance-ment, as if insatiable consumption and haphazard expansion represented economic growth, as if riches were defined only as money, made and lost in a moment, not as natural treasures it took Earth all of evolution to produce. Surely there are ways to enhance the quality of life without de-meaning life at the same time. The truth is that no unyielding economic or technological imperative is forcing us into the wholesale destruction of Earth's riches. We are cursed, not by fate, but by ourselves. We have allowed ourselves to become resigned to our present-day lack of clean water, lack of resources, lack of unmined, unlumbered, uncorrupted vistas; yet the fact remains that these luxuries of life existed once, in plenty, and continue to exist, to a fragile and greatly reduced extent, still. If we must give a reason for protecting life and its habitat from saccage, then let it be this: Protect them because somehow, some way, in an infinite cosmos, these glories appeared on Earth. Protect them because they are here.

* * *

Saccage begins where life began: in the nurseries of the sea. Life thrives in three parts of the ocean: the surface waters, penetrated by sunlight, where plant life blooms; the bottom, where organic detritus settles; and the area in which these two life-nurturing factors combine, the continental shelf, which begins at water's edge and extends out to depths of about six hundred feet. As I've stated before, the most fragile part of the shelf is also its most fertile, the shallow coastal waters. Sea life gathers at the coastline as land life gathers at the river; there fish feed on flora and drop their eggs in seafloor prairies of Posidonia. These waters—just one half of 1 percent of the total ocean space—support 90 percent of all marine life. The world's coastlines, I once calculated, have fewer square acres than the world's rivers.

It is precisely on these scant coastlines that *saccageurs* wreck their havoc. Construction companies dredge for sand and gravel, scraping away Posidonia and fish hatchlings, shaving the bottom like a bald man's pate, leaving a record of their ruination in catastrophes like the collapse of the once-thriving herring fisheries of the English Channel.

More hatcheries succumb when construction workers then dump their dredge spoils and build landfill extensions in coastal waters: What act, after all, is more inimical to life than burial? In 1979 on France's Côte d'Azur we filmed one of the most heedless and ultimately pointless such projects I'd ever encountered. Our cameras recorded bulldozers advancing like tanks toward the Mediterranean, dumping loads of rock into the sea alongside the airport at Nice, extending the existing landfill to make way for an airstrip. Construction workers scooped the land from a hillside, the crest of which was crowned with the home of an old man who had lived there all his life. Like many who battle eviction, he refused to budge; if his hill succumbed, he would succumb with it. The bureaucrats proceeded, isolating him and his house, where he continued to reign perched atop a cone of dirt.

Yet the saccageurs, in their desire to expand the airport—to bring in more tourists, more business, more profits—were overzealous. They heaved their masses of rock and dirt into the ocean without first building a retaining wall. They thus created a precarious underwater cliff that then collapsed. The resulting avalanche gave rise to a tidal wave that ripped toward shore, hit the coast, and then ricocheted back out to the sea. In the quarter-hour rage of water, the wall of water hit the coastline from Cap Ferrat to Antibes. Windows shattered. Private homes sustained millions of dollars of damage. Boats vanished. A woman was swept out to sea, never to be seen again. At the airstrip, three men forever disappeared and six men were killed. Public outrage welled up like the tidal wave itself. Work on the airstrip was abandoned. Months later, I attended a CIESM meeting in Sardinia, where geologists reported that a turbidity current—a flow of liquid surging with dirt and rock, heavier than water—had raced across the bottom of the Mediterranean, spitting out the debris it carried from the decimated airstrip. The geologists showed me a photograph of a sunken bulldozer that the current had thrown halfway to Corsica. In the end, the only victor appeared to be the little old man sitting up on his unconquered crest, king of the hill.

A few years later, farther up the coastline, I became involved as engineers once again planned another assault of saccage—on the Île de Ré, a little island whose bracing salt air and singular golden light have attracted numerous photographers and landscape artists. Ile de Ré's charm attracted tourists as well; its population of 10,000 year-round residents swelled to 200,000 in the summer. Day-trippers, queuing in their cars for the return boat trip, waited hours in a bumper-to-bumper line for the ferry. The general council of the area decided to build a bridge joining the island to the mainland.

Newspapers soon headlined a public debate as both government agencies and private citizens warned local officials about the ultimate damage their project would do. The island's fragile dunes were already compromised by huge parking lots. Tourism, they said, should be limited rather than encouraged. An environmental impact study, required by French law, concluded that a bridge would not serve the public interest. The local prefect barred construction.

When bridge proponents insisted that another study be done, the staff of my French environmental society, l'Équipe Cousteau, and I launched a study of our own. We proposed some rational alternatives— streamlining and increasing ferry services, offering free minibuses on the island along with a monorail—all of which would have been less expensive and less destructive than a bridge. Again the prefect barred bridge construction. Again bridge proponents appealed the decision, which went back to the jurisprudence system for yet another report.

To our astonishment, before the report was even concluded, construction crews began building a bridge! In Paris, I outfitted my entire staff in overalls and hard hats, rented cement mixers, and appeared on the banks of the Seine to lay foundation stones for a fictitious bridge to join the Parisian Right Bank to the Left. I'd arranged for a fictitious policeman to interrogate me; I gave the answers that the Île de Ré bridge builders had offered: Permit? Not necessary. Benefit of the bridge for Paris? Nothing. We received a great deal of publicity but achieved nothing as far as results. The final Île de Ré report, issued by the state council, was everything we'd hoped it would be: It required all public authorities to stop construction of the bridge. Yet by the time the report was published, the bridge had been built.

A few years later, the president of the Friends of Île de Ré group described how tourists' numbers had burgeoned and how stores were multiplying on a no-longer-so-charming commercial strip. He lamented that the identity of Île de Ré would be different in a decade. In my view,

Île de Ré's identity was already different. Île de Ré was connected to the mainland by a bridge. It had become a cape. It was no longer, and never again would be, an island.

Even more ravenous than saccage against life's habitats is saccage against life itself. Hunting for pleasure persisted when it was the sport of kings, when only royals were permitted in royal forests. In hunting terms, democracy did not reduce the king to commoner but instead made all commoners kings. Natural selection—which suppresses the weakest animals—was replaced by throngs of hunters who strived to kill the strongest, sleekest, healthiest.

The hunter with his gun—like a diver with a knitting needle—is not necessarily motivated by greed; he can be merely misguided by a lack of information. When we filmed the North Country forests of Canada, we searched the woods and found no wildlife to record: Gone were the caribou, the reindeer, the foxes, and the wolves. The forest itself was receding by more than thirty miles a year. One day I met the president of the Saskatchewan hunters' association. He was indignant when he heard my objections to hunting. "Hunters are the friends of nature!" he protested. "We protect wildlife; it's advisable to thin out herds!" I answered: "Here is our helicopter. Climb aboard." Together we flew over the forest. By the time we landed, he was livid—but no longer with me. "It's a disaster," he commented bleakly. He'd seen that the herds had not been thinned; they'd been exhausted. In order to understand, he had needed the helicopter as I had needed an Aqualung. He despaired at the empty forests as I had despaired at the empty seas.

Saccage against life is often even more indifferent. In the Amazon rain forest, we filmed poachers who by night stalked the legally protected caiman, a reptile similar to an alligator, finding the creatures by beaming flashlights through the jungle and looking for the telltale double gleams reflected by their eyes. With a rifle, harpoon, and knife, one poacher can kill and skin 40 caiman in one night. Poachers killed an estimated 40,000 caiman a month in the Pantanal. The 40,000 carcasses were left to rot. The poachers were interested in nothing more than the animals' flank skin. They killed one animal to obtain the leather for one shoe. Environmental groups raised a parallel issue when they protested the clubbing of baby harp seals in Canada and refused to consider the government's explanation that seal populations had grown beyond natural support levels. The debate about cruelty to animals can, and probably will, continue to the end of human time. Still, no one can deny the plain and simple sin of waste. When I filmed the Canadian harp seal hunt, workers swiftly skinned the animals, snatched the furs, and departed—leaving the beaches behind them littered with squandered meat left to putrefy.

Those who hunt the sea are also governed by havoc and waste rather than by need and efficiency. There was a time when the fish that escaped nets and lines could take refuge, and gain time to replenish their numbers, in rocky crevices; no more. Spearfishermen wipe out groupers and other sedentary fish from rocky shelters; they terrorize fish during spawning seasons. Despite the fact that the actual figures for a spearfisherman's take are small, one aggressive man armed with a gun disturbs the behavior and reproduction of fish to a degree to which only those who have dived below and seen the results can testify. I've witnessed the destruction of an entire reef community in one week by a few spearfishermen. In the Medas Islands, I once dived among a multitude of beautiful daurades; twenty-five years later I returned to find a nearly empty shooting gallery, still frequented by unrelenting daily squadrons of spearfishermen.

Even less rational, even more rapacious, are the saccageurs who compound one abuse by adding another, who suffer from what I'd describe as the scapegoat complex. They wipe out whole populations of targeted creatures, blame the carnage on natural predators, and then kill them too. Over and over I've seen this domino effect of saccage. Just after World War II, while I was still a commissioned officer with the French Navy, we sailed to the islands of La Galite, in Tunisia. There we often saw monk seals, the only marine mammal indigenous to the Mediterranean, playing in the water. But even then lobster fishermen were already bitter. They'd seen monk seals surfacing with lobster pots on their heads. The animals were eating lobsters that "belonged" to humans. When the lobster business became industrialized, the commercial fishermen got guns. By the time we returned to La Galite in 1977, the fishermen had reduced the seals, among the world's rarest mammals, to numbers pitifully close to no return. By overfishing they wiped out the lobster fishery as well, but by then they had no monk seals to blame.

When we filmed sea otters in Morro Bay, on the California coast, we found that abalone fishermen suffering from the scapegoat complex were dismantling an entire ecosystem. They bitterly complained that otters ate "their" abalone catch. The angry fishermen stalked the otters; we even found the remains of one otter with a smashed skull. Later, the star of our documentary, an otter that had frolicked in the waters and even teased my son Philippe, washed ashore riddled with shotgun wounds. The abalone fishermen had failed to piece together nature's puzzle: Otters, granted, eat abalone; but they also protect the abalone nurseries by eating sea urchins, which devour the kelp gardens in which abalone lay their eggs. As fishermen killed otters, the kelp-garden nurseries of abalone eggs lost their protection from the famished urchins. The multiplying urchin populations reduced the kelp gardens, and with them, the abalone nurseries. Having unwittingly saccaged the abalone nurseries, the abalone fishermen—looking for a scapegoat—focused on their next target: the sea urchins, of course! The fishermen attacked the urchins with hammers, smashing the creatures—and thus releasing sperm that promptly increased the urchin population further and wiped out the abalone beds on which the fishery depended.

*　*　*

As they blame one scapegoat after another—looking for revenge on seals, otters, urchins—rarely do the saccageurs look in a mirror; rarely do they blame themselves. Nor do they stop to consider simpler solutions. I could only shake my head in bewilderment when we walked the abandoned fishing villages of Canada. Trout populations had nearly disappeared in Lake Superior and in fact had vanished from Lakes Huron and Michigan. The trout's decline coincided with two events: decades of overfishing by humans and decades of increasing numbers of lamprey fish, a natural predator of the trout. The fishermen, of course, blamed the fish, and promptly dumped a poison that, if used with utmost care, might spare remaining trout while wiping out the lamprey larvae. Even my aunt Almaid could have scolded the Great Lakes scientists for adding one abuse unnecessarily onto another. When I was young in the village of St.-André-de-Cubzac, we eagerly awaited Sunday—the day that Aunt Almaid proudly prepared her *lamproie à la bordelaise* in her country kitchen. She piled vine shoots high on the hearth, then over a blazing fire she hung a cauldron in which she boiled pieces of lamprey. For hours she'd add vine shoots to the fire and test the broth with a wooden spoon. At last she'd extend a skewer into the flames until it glowed red hot, poke it into a lump of sugar, and spear the lamprey meat with the resulting caramel. A splash of red Bordeaux was her final touch. A far more creative solution to a lamprey problem, for my tastes, than poisoning the Great Lakes.

As they plunder for profit, those who loot the lands and seas think only of the cash in their pocket today; they even pillage their own catch of tomorrow. Nylon nets have revolutionized the fishing industry, say the fleet owners. Yet those who use nylon nets have also created a competing industry—ghost fishing. Hundreds of millions of pounds of plastic gear from commercial fleets—nets, lines, buoys—sink to the bottom each year. Cotton nets disintegrate; nylon ones endure. In Santa Barbara we filmed an all-too-common side effect: a seriously wounded sea lion, his neck slashed by a nylon net that had encircled his throat and cut through skin and muscle as he grew. As I dived between Tunisia and Italy, in heavily fished waters, I saw lost lines, nets, traps draped on rocks along the bottom, as well as steel lobster pots that replaced the old wicker baskets—all of them trapping lobsters and fish without the hand, or benefit, of human beings. Fishermen above remained oblivious to the black holes of the sea universe, the ever-present nets and traps that attract and swallow up the life below.

The most haunting experience we had with squandered sea life came to us as we filmed the wreckage of squandered human life, of the gruesome toll of wartime. My son Philippe and I had sailed the Central Pacific into a group of volcanic islands in the heart of Micronesia. There we anchored in Truk Lagoon. Long ago a world-renowned harbor, Truk had become, in the course of a single U.S. bombing mission in

World War II, a sunken junkyard, awash with the wreckage of an entire Japanese fleet. Day after day we dived, exploring dozens of ships.

But where were the large fish?

The team filmed the sunken silhouettes of war, anti-aircraft guns pointing toward the surface, monuments to sailors who had once fought to the last.

Where were the groupers, the moral eels, the lobsters that everywhere else inhabit the skeletal frames of sunken ships?

Philippe could not resist the temptation, despite the difficulties represented by a 280-foot depth, of diving to one last, almost-out-of-reach warship. Fighting the illusions of Rapture of the Deep, Philippe descended to the spectral structure and forced open the door of a watertight compartment. And then he beheld the scene he called his vision of hell. Skeletons, detached bones, leering skulls, preserved in the oxygen-void waters, a testament of war delivered to the living from the dead.

Somehow they spoke to him not just of the saccage of wartime but also of the saccage of peace—the war that had been declared on the creatures of the lagoon. Philippe had seen vast hordes of canned food and drink below—but where were the stocks of ammunition? Plundered! That was it: We had seen no large fish—none of the fish that take decades to grow—because natives had looted the ships of their explosives, used them for dynamite fishing, and thereby looted the sea.

Around the world we've dived in waters that resound with the distant bursts of dynamite explosions, the sonorous cadence to which the saccageurs destroy sea life. They blast apart flora and fauna, squandering 90 percent of the fish they kill—which sink to rot on the bottom—in order to grab the limited booty that rises to the surface in a bloated circle of death.

In the Philippines, we saw fishermen who even involved their children in the carnage, sending youngsters down to pack dynamite into crevices of coral—implicating generations of the future in the destruction of evolution's creations of the past. A major source of the world's tropical aquarium fish, the Philippines officially bans another form of saccage against coral and its inhabitants: poison fishing, the nonetheless most widely practiced form of fishing in the Philippines. Parents and children squirt sodium cyanide into the reef, hoping to anesthetize tropical fish so that they can be gathered up unscathed. The cyanide in turn kills off sequential portions of the reef as well as causes liver damage to many of the fish—an ominous portent for the future market, but all the better for immediate sales, because large percentages of the fish die rapidly after purchase. Authorities fight a losing battle trying to patrol their coasts, helpless to outwit natives who depend on their pitiful catch for their meager livelihoods. No one suggests that more-efficient fishing methods, with which fishermen could reap all that they kill, rather than just 10 percent, might augment their catch, increase the fish stocks, and decrease damage to the reef on which the fish depend.

Neither the hungry Filipino fishermen nor the homeless Maldivians should shoulder all the blame for the world's assault on coral reefs. The covetous desire for coral has nagged human beings since antiquity and certainly explains why the rarest forms were long ago wrested from the sea. Black coral gained its allure through its rarity—appearing only in forbidding depths, taking centuries to grow, rising into ebony-colored trees, with massive trunks and branches that are difficult to harvest in the deep. Although Muslims treasure pieces, polished to a shine, as prayer beads, the looting of the precious trees from the reef around Belize, the second-largest fringing reef in the world, has long been strictly prohibited. One day in the fifties, we were sailing off the area when a small boat approached. The disheveled men aboard, eyes darting, asked us— then begged us—to refill their air tanks. For what were they diving? Why were they so desperate for more of it? They threw open their hold; it was brimming with black coral. They were pirates, thieving the invaluable branches from prohibited areas.

I learned the wages of the sin of saccage by committing it, unintentionally, myself in Belize. One day we prepared to set sail. *Calypso* would not budge. Our anchor was stuck, wedged tight into a narrow crevice of black coral. I sent divers below with crowbars to pry us loose; still we could not extract the anchor. We hooked the anchor chain to our winch and gave a mighty pull—up came the anchor. Off cracked an enormous piece of black coral, a boulder that went tumbling down into the depths.

When I've told the story, in remorse, to some old sea salts, they've consoled me with the remark that a piece that size would not be missed for the generation or two the gash would take to grow back in. But what of all the ships and boats doing the same thing? I instantly called an end to anchoring *Calypso* carelessly. The worldwide problem could be similarly simple to correct by requiring that crews pass an ecological exam. Yet around the world, millions of little pleasure boats, yachts, and ships daily drop anchor and break off coral from the world's reefs. Hordes of tourists who don't threaten the reefs in boats destroy them by patronizing souvenir stores and museum shops; they purchase broken pieces of coral and the sometimes primitive shells they contain, a commercial demand that drives teams of Tahitian divers alone to destroy an average of more than six miles of reef a week. In 1979, more than twenty years after our encounter with the brigands from Belize, we took our diving saucer down into waters off Corsica to see red corals out of reach of the voracious divers in the area, who had stripped the precious jewelers' coral from all but prohibitive depths and even then were daring the dangers of Rapture to hack away for more—soon to be cut by lapidarians into objets d'art. We despoil the coral reef—womb of life that regenerates world waters—for baubles and trinkets.

Not satiated with saccaging the coral reefs of the sea, we destroy in turn what can be considered the coral reefs of land: our rain forests. A simple spin of a schoolhouse globe testifies to the costs of destruction: Our Sahara stretched out, six thousand years ago, as one of the densest woodlands on the planet. Abrupt climate change with little input from humankind devastated Saharan civilization. What will happen now that human beings contribute to woodland destruction? We render ourselves deaf to the question and continue to raze world forests at the sense-defying rate of 150,000 square kilometers a year. The poorest people of our global village need wood to burn; instead of expending money to help keep them warm, we expend our earth, letting the forests disappear and losing, in the flames, the undiscovered natural medicines, the undiscovered kinds of seeds, the unexamined resources of the rain forest. While sands inch across the face of once verdant acres, we lose evolution's bounty as well, with between one half and two million plant and animal species—perhaps 20 percent of all species on Earth—expected to vanish by the year 2000. The weary Earth can replace them, of course—at the same rate she creates them, about one new species every million years. Perhaps the replacement life will appear at a time when the human species has evolved to the point of deserving it.

For now, we're trapped in saccage's vicious circle: The degradations we ourselves have caused dull the senses and thus become self-perpetuating. With each extinction of a species, each obliteration of a land mass, each defilement of a body of water, people come to realize that life is less pleasant, but then they discover that desolation is not death, that they can survive, and they finally accommodate by adjusting their lives downward. Those who enjoyed a wild, natural vista in their youth now tell their children they remember the place before it was a parking lot. Life gets worse, but at least it goes on. The impoverishment of the environment parallels the impoverishment of the spirit. Each progressively causes, then aggravates, the other. Drab planet, drab lives. Depleted waters, depleted lands, depleted souls.

During my team's two-year mission to the Amazon rain forest, my son Jean-Michel visited the Jivaro-Achuaras, a native tribe living deep within the jungle near the Peru-Ecuador frontier. He became close friends with the chief, Kukush. One day Kukush informed him that the wisest members of the tribe had met; they needed a new canoe and had discussed the grave decision of whether or not to fell a tree. Finally, they had decided to proceed and prayed to the gods to forgive them.

But to Kukush, the prayer alone was not sufficient. To replace the tree that they had taken, Kukush planted several hundred more. They were saplings; they would never grow to serve him or his tribe within his lifetime. These trees were for his grandchildren. They were for Earth.

CHAPTER SEVEN

CATCH AS CATCH CAN

T HE old Jamaican I saw, bobbing alone in his ramshackle skiff, embodied Hemingway's luckless fisherman, wearily persevering in his struggle with the sea. His skin had shriveled like the peel of a fruit, baked for a lifetime of some seventy years under the glare of the Caribbean sun. Time had spared his weatherworn body no more than his ragged boat. He worked with a wiry strength, hoisting large fish traps in and out of the craft, but his wasted form was that of a hungry man. Bones jutted from his rib cage, and wizened cheeks draped the hollows of his face like the skin of a battered drum.

I first spotted him as I returned in my Zodiac from a dive off Jamaica's northwest coastline. We had been exploring Montego Bay's spectacular reefs, filming the shapes and shades of coral that bloomed among the labyrinths below. It had not taken long to realize we were swimming over a splendid desert: Though festooned with worms, urchins, and snails, the reefs were curiously devoid of the profusion of fish that normally thrive among corals. Fishermen know their waters well; and so, when I saw the old man row in and pull his boat ashore, I beached my Zodiac and wandered over to talk.

His name was Moses and he'd fished the bay for more than half a century. Had the fish always been so scarce? I asked. "Oh, no." Moses gestured toward his gear. "I used to catch large fish in my traps. But now I see no more big fish."

I knew that Jamaica, which had never been able to eke a sizable catch from its craggy-bottomed seas—poor trawling grounds—had nonetheless somehow doubled its take in just ten years. I'd heard that increased fishing pressure had overexploited the main commercial species, like snappers and groupers. And I could see for myself, even at that moment, fishermen far offshore, competing for migrating fish in the open sea with old boats they'd newly motorized. Those who had nothing to spend other than effort, like Moses, had over the years resorted to using smaller and smaller mesh in their traps, catching smaller and smaller fish. I looked down at the traps that Moses had rigged out of chicken wire. The size of the mesh was a mere half inch.

"With such tight mesh, you give fish no chance to grow. You will soon catch nothing at all," I said. Moses's gaunt face remained expressionless. With gnarled fingers he picked his catch from the traps: a dozen two-inch reef fish, more ornamental than edible—a whole day's work for two mouthfuls. Finally he looked up. His sunken eyes flashed with brief but piercing sadness. "I know what I do is not good for the years to come." The old man scooped his pathetic catch in the palm of his hand. "But I must eat."

Moses's tragedy has rapidly widened into a global dilemma. Around the world, more and more boats chase fewer and fewer fish. Not only have mediocre fishing grounds like those off Jamaica been reduced; the richest fishing grounds on Earth have, one after another, eroded into the fishless deserts Hemingway long ago imagined. People who depend on the sea for their livelihood—or, like Moses, for their lives—are not the villains but the victims, brushed aside, mere expendables in an international fishing regime where anarchy reigns. Five-thousand-ton ships, armed with advanced technology to hunt down their take, vacuum most fish from a school in one or two sweeps up a coastline; with their third pass, they leave waters nearly empty. Commercial fishing industries honor a creed of Catch as Catch Can, deeming it more expedient to eradicate an entire fishing ground, investing profits for immediate returns, than to nurture fish as a fragile but renewable resource.

In this dead heat for the last fish, industrialized fleets take the biggest catch, but no one wins. Our voracious fishing depletes rather than replenishes the world's protein coffers. Pigs and chickens in wealthy countries feast on more than one third of the world's catch, now of such small size or low quality that it is ground into livestock feed. The people of those wealthy countries derive only minuscule fractions of their protein directly from fish; humans in poor nations draw even pitifully less sorely needed nourishment from the sea. Nor does fishing contribute significantly to the world economy; the social waste in overequipped fleets, not to mention the unemployment and bankruptcies that punctuate booms and busts of fishing collapses, drains nations of net income, while only individual corporations benefit.

Moses cannot conserve for tomorrow, or he would starve today. The world fishing community need not be forced into his short-term choice. Yet while even Moses recognizes, and agonizes, about overfishing's consequences, the rest of the world apparently dismisses them. Confronted by the need to manage migrating world fish stocks on a cooperative global level, individual countries instead have declared two-hundred-mile coastal domains, cutting world waters to national pieces, creating a Law of the Sea that simply issues each nation a permit to fish lawlessly in its "own" sea. Confronted by manipulative governments and emasculated by a lack of authority, international regulatory agencies indulge fishing industries rather than control them. Confronted by the need to amortize their expensively equipped ships, thus addicted to ongoing bank loans and government subsidies, the fishing industry itself serves only to perpetuate the chaos. Together these special interests speak out for reduced regulations and increased price supports, for more government loans, more subsidized boats, higher technology. Together they speak of even more monumental catches as they make empty promises about bounty in the years to come. Few speak for the consumer, who now pays exorbitantly for fish in taxes as well as in spiraling grocery bills. Few speak for the fish populations, as additional stocks of edible species continue to dwindle. Few speak for the seas, as consecutively depleted coastlines merge into one depleted ocean. Everyone speaks for the present. Few speak for the future. Everyone speaks for themselves. And almost no one speaks for Moses.

Not so long ago, great expectations about the ocean's abundance— visions of the seas as a brimming bargain basement—seemed warranted.

As recently as the turn of the twentieth century, Thomas Huxley rhapsodized about the "inexhaustibility" of the seas, and for decades the soaring world catch appeared to justify his proclamations. By 1950 fishermen were landing 21 million tons of fish each year; their take multiplied another two times by 1960; it soared to a total of 70 million tons of fish by 1970. Exultant fishing-agency officials envisioned a catch, for 1985, of some 89 million tons. Someday—who knew—maybe 290 million tons, even 580 million tons!

All these illusions shattered when skyrocketing catch statistics abruptly collided with a ceiling. The enthusiastic projections for 1985 fell short by some 15 million tons. The speeding rate at which the world catch had been increasing had ground to a near halt, and Western Europe's annual catch actually dropped. Facing such evidence that the ocean does indeed have a bottom, world fishing agencies deftly backstepped, moderating their projections. Still they incorrigibly predicted that fishermen could somehow land 120 million tons—a figure that presupposed we could nearly double the catch of fish for food—and some said this could be done even by the year 2000.

A closer look at the figures, however, suggests that the outlook for world fisheries is far more grim. Unfortunately, officials based their predictions on a total world catch statistic that is misleading; in the fishing industry, even "straight facts" can be dangerously deceptive.

First, by including vast tonnages of "unconventional" species—the euphemism for "trash" fish—in more recent totals, authorities camouflaged collapses of the major commercial fish stock. The round-nosed grenadine, for example—hardly a popular delicacy—was for years not even specified in North Atlantic catch statistics; within twenty years of its inclusion in the books, it was magically adding 30,000 tons to the grand total.

Buried under such tonnages lies evidence of stock failures that have continually and successively circled the globe: Haddock have been depleted off Canada and the United States; herring, off Norway and Japan. Halibut have disappeared from New England and Greenland. Flounder are diminishing on the Great Banks; lobster, in the west Atlantic; salmon, off Japan, the former Soviet Union, and Alaska; shrimp, off India, Indonesia, and Alaska. And throughout the waters of the world, cod, hake, sardine, anchovy, pilchard, mackerel, and yellowfin tuna falter in the worst condition ever.

The figure for total world catch is misleading in yet another way: It obscures the fact that, though we are barely maintaining our harvest, we have had to greatly intensify our efforts to get it. Between 1971 and 1983, for example, the workforce of American fishermen swelled from 136,500 to 223,000; Americans embarked on a building boom as well, expanding their fleet of industrial fishing ships, five net tons and over, from 13,100 to 21,100. Yet while the numbers of fishermen and ships nearly doubled, the total U.S. catch of fish for food actually dropped below the tonnage caught back in the fifties.

Nor does the present world catch reflect the exponentially intensified technological powers bestowed on fishing fleets. A few decades ago, developed nations began to pour funds into huge "factory" ships— veritable floating industries, equipped with processing machinery to behead, bone, fillet, and freeze virtually generic blocks of unrecognizable fish for storage. These ships need not return to shore even to deposit their take as they forge on, for months at a time, in a relentless pursuit of more catch. Huge trawlers drag the ocean floor with conical bags of nets, held open by two-and-a-half-ton steel gates, scraping up fish from the bottom as they destroy the habitat. Slowly cruising long-liners daily let out more than fifty miles of line, held up by buoys; about every two hundred feet along this floating main line, "dropper" lines, with bait and hook, dip down to snag the large, swift creatures of the deep.

Accompanying the giant long-liners are fleets of tiny "catcher vessels." Dark, dank, cramped, and grimy, they swarm with workers who toil in alternating shifts, through day then through night, incessantly seizing take from the lines. Though preceded by their notoriety, a catcher crew nonetheless stupefied me when I first encountered one in the fifties. *Calypso* had sailed the Red Sea through the straits of Bab el-Mandeb, across the Sea of Oman. Heading toward the Seychelles Islands, in the Indian Ocean, we had slipped over waters so serene that the sea, glittering with bioluminescent salps, fused with the starlit sky; our wake seemed to tear the fabric of an inverted aurora borealis. Out of this misty calm the next morning clattered a rusting Japanese catcher vessel, smaller than *Calypso*, choking with laborers. The workers moved as though they were half men, half cats, scrambling over rigging, prowling the decks. Some slinked into a rowboat and paddled to *Calypso*. Almost in an instant, lithe young bodies swarmed over our ship, peering through every porthole, poking through every opening. Our invaders understood no French or English; their sole communication with us consisted of identical grins plastered across every face. The stench of fish settled over our ship. Just as I was recovering from my surprise, the agile, dark shapes vanished. Their curiosity sated, the cat-men had crept back into their rowboat and paddled off to toil again in the confines of their grim little mobile sweatshop.

Totally intrigued, I decided to return their visit. But on board their sailing slaughterhouse, there was no room for wandering, no time for us or our pantomimed questions. Crew quarters spilled out inmates; life on board was nothing more than incarceration. That old tub was overcrowded with workers, overloaded with fish. Silvery carcasses of tuna, marlin, sharks, and sailfish rained down on us from all directions, as fast as the laborers could reel in dropper lines, unhook them, rebait them, and return them to sea.

Back on *Calypso*, we followed the buoys that supported their main line, sailing for twenty miles before we could even make out the hulking silhouette of their factory ship, surrounded by a half dozen more catchers like the one we had visited. About that time, Japan's entire long-liner fleet paid out enough line in just one year to circle the globe five hundred times, setting four hundred million baited hooks. Even today, when *Calypso* crosses the Pacific, Atlantic, or Indian Ocean on any latitude, wherever we go, our screws cut through dozens of longlines, entangling the oceans like a global spiderweb. More startling than the total tonnage of fish these ships land, however, is the tonnage they fail to land. Between 1957 and 1977, the total world catch of bigeye tuna rose some 270 percent; yet the Japanese catch of bigeye tuna *per hook* fell almost by half.

Further technological prowess has only aggravated the overexploitation. Using even-more-efficient giant "purse seiner" ships, fishermen surround vast clouds of fish with a three-quarter-mile, fifteen-ton net; by pulling a drawstring around the bottom of the net and shutting it tight, they wipe out an entire school at a time. The mammoth purse seiners rely, in addition, on satellites that monitor sea-surface temperatures and thereby predict the paths migrating fish will take; on sonar, which accurately targets fish schools; even on computerized video plotters that precisely position the gargantuan nets—leaving schooling fish with no chance whatsoever. While the use of such advanced electronics has augmented the catch in the past, it is exhausting the seas for the future. When technology doubled the efficiency of the menhaden fleet off the U.S. east coast, the annual catch actually dropped all the way from 712,000 metric tons to 161,000. North Sea herring supported a fishing industry for two centuries; two years after Norway enlarged its North Sea purse seiner fleet from 16 vessels to 326, the spring-spawning herring stock totally collapsed. When purse seiners targeted the Peruvian anchovy, the catch soared from virtually nothing to more than 12 million tons—one fifth of all the fish harvested annually around the world. The huge anchovy fleet pulled in an average tonnage each day that rivaled the catch the entire U.S. tuna fleet at the time caught in twelve months. Three years after the peak anchovy catch, the greatest fish stock on the face of the earth collapsed.

Fishermen, of course, have obliterated fish stocks without the aid of technology, long before industrial fishing weighted their boats with panoramic sonar recorders and flasher-style fish-finders. Advanced technology has only added speed to devastation; while once it took decades to ruin a stock, today we can accomplish the feat in only a few seasons.

Some eleven thousand years ago primitive tribes slowly began to comprehend that they could gain more food by planting seeds and raising animals, by renewing their supplies of food rather than by hunting them out. They relinquished the sticks and stones they had used to knock down birds; they renounced the nomadic life in which they had advanced from one depleted area to another. They settled, farmed, and founded civilization. One hundred and ten centuries later, modern corporate fishing industries still lumber down to the sea to hunt for fish.

While Neopaleolithic man perceived the dangers of depletion, the modern industrial fisherman remains literally blind to the lesson; for he is the only huntsman who cannot see his quarry.

If he could see—if the water world were as visible as vast tracts of land—the fisherman would lay eyes on a sobering sight. Most of Huxley's "inexhaustible" sea—the vast expanses of water in the open ocean—is at best a biological desert. Only a small portion of the sea can produce phytoplankton, the basic food on which all ocean life depends. Phytoplankton pastures bloom, and fish grazing on them proliferate, only in the few corners of the sea where sun penetrates and nutrients abound—shallow continental shelves, island wakes, sheer underwater cliffs against which rare upwellings spew their rich bottom broth up toward the sunny surfaces. An acre of even these fertile waters produces a smaller crop of plant life than an acre of land; thereafter, each ascending stage of sea life, from plant to more complex animal, is smaller still. The cod, salmon, and flounder that are prizes for fishermen occupy the upper reaches of the food chain; tuna preside at the very top. While it takes only 15 pounds of grass to produce 1 pound of beef, the ocean must supply some 1,000 pounds of plant life—so relatively rare in the sea—to feed all the creatures that will ultimately feed 1 pound of tuna.

Into this fragile marine world forges the modern fishing industry, with its advanced technology and primitive mentality. It chases the sea's scant creatures with light airplanes, helicopters, satellites. No hunter would track rabbits with radar; yet, in the sea, industrial armada use the equivalent of armored tanks and automatic weapons to eliminate a population of squirrels.

For what?

Fish does not now—nor can it ever—feed the world.

Fish provides on average only 5 percent of all protein humans consume daily. Many developed, wealthy nations draw even less of their protein from the sea: Americans obtain just 2.2 percent of their protein from fish; French, 3.9 percent; Italians, 2.8 percent. Even nations that exceed the average—notably Spain (10.6 percent) and Japan (17.8 percent)—nonetheless dine on seafood more for taste than for necessity. Among the developed nations reporting food consumption to the Organisation for

Economic Cooperation and Development, not one would on average incur any protein deficiency even if fish were completely removed from its dinner tables; the French, Spanish, British, and Americans, in fact, are so surfeited with protein that, were they to forgo fish entirely, their remaining average diet would still provide twice the protein necessary for good health. Despite Japan's insistent protestations that any restrictions on fishing would jeopardize its food supplies, the average Japanese eats 100 percent of the daily protein requirement, as well as another 68 percent to spare, before even a gram of his fish consumption is tallied into the count.

Many agencies try to justify intensified fishing efforts by saying that fishing "feeds" poor nations to a far greater extent than wealthy nations. Is this true? Not really.

Granted, no one can argue with the facts officials cite. "Four out of ten people in the Third World, people who live in a narrow zone between bare adequacy and starvation, get 30 percent of their total animal protein in the form of fish. In some nations the proportion is 70 percent," an assistant director general from the U.N.'s Food and Agricultural Organization (FAO) recently stated. "Fish is the meat of the Third World."

Others chime in with a litany of figures filled with apparent hope. Fish provides more than 50 percent of the animal protein obtained by the people of Chad, the Ivory Coast, Jamaica, the Republic of Korea, Malaysia, Mali, Senegal, and Uganda; those in the Congo, Vietnam, and Indonesia depend on fish for as much as 60 percent of their animal protein intake.

How do these impressive declarations exploit the plight of the poor? The tragic truth lies on their plates. In fact, fish provides so large a share of protein in the diet of starving people not because they have a great deal of fish to eat, but because they have nothing else to eat. Their total intake of fish, necessary for their survival, amounts to even less than the portions of fish people in advanced nations eat in excess of their protein needs. Yes, the Indonesian depends on fish for more than 50 percent of his animal protein. But that grand percentage translates into only 9 pounds of fish for an entire year, while the average Japanese feasts on 71 pounds of fish annually, the Russian on 23 pounds, and the American on 12 pounds—as well as 117 pounds of beef.

The argument that we must deplete the seas in order to feed the starving—when fish are eaten by the overfed and the starving go unfed—seems all the more empty when you take a glance at the rest of the food on the Third World platter. Scientists estimate that an average adult man needs 41 grams of protein to replace tissues that break down each day. Inhabitants of the Third World average only 11.5 grams of animal protein daily—just 2.4 grams of which come from fish. If they don't obtain the rest from vegetables and grain—or if, with nothing else to eat, their bodies cannot use fish as protein to rebuild lost tissue, but must instead burn it for energy—they will literally wither away, developing the protein-deficiency disease kwashiorkor. Many will die.

This in fact is the fate of many from those populations extolled by fishing lobbies as the world's primary beneficiaries of fish for food. The people of Jamaica—supposedly big fish eaters who obtain more than half their protein from fish—number among the nations whose children under the age of four have been hit most severely with the scourge of protein-energy malnutrition. While fishing agency officials boast that one dozen African nations south of the Sahara derive much of their protein from fish, some eighty-three million Africans suffer from malnourishment. Each year, five million African children die of starvation.

Whatever the glowing percentages, whatever the optimistic figures, whatever the sleight-of-hand statistics, fish aren't feeding the Third World, because the Third World remains unfed. It does not well recommend fishing as a food supply to call it the cornucopia of the dead. The 50 percent statistic for fish protein consumption in certain countries is a statistic of shame. It is 50 percent of not enough.

Yet not only does the fishing industry fail to feed the hungry; accumulating evidence indicates that the fishing industry does not even seriously *intend* to feed the hungry.

If we really wanted to feed the world with fish, then why do we throw so much fish away? Officials say that each year, fishermen toss some 5 million tons of fish—more than 550 tons an hour—back into the sea to make room on their ships for the catch that will bring them the highest prices. Extraordinarily wasteful trawler and purse seine nets do not discriminate between fish the industry wants and those it doesn't; they scoop up them all. When *Calypso* sailed south of Newfoundland, we followed the *Croix de Lorraine*, a large trawler dragging its net over the Great Banks. Undersized and undesired fish filled half its bulging net; the trawler crew removed the "commercial" fish, then unceremoniously heaved overboard those that would not bring enough profits. Almost all the discarded fish die; jettisoning them usually benefits only the scavenger seagulls that pluck flesh from the waves before it sinks to the bottom of the sea. But that day some other scavengers descended with the gulls: *Calypso* divers jumped into a Zodiac and scurried about the water to recover a hundred pounds of good-sized red scorpion fish, from which our cook prepared a delicious traditional bouillabaisse.

If we really wanted to feed the world with fish, then why have we fed so much of our fish to livestock? Pigs and chickens eat more than one third of the world catch. Some countries feed their farm animals an even greater portion of their fish: In the United States, where the average American consumes twelve pounds of fish a year, another fifty-two pounds per person are annually fed to poultry and swine—a share greater than twenty times the amount the average inhabitant of India eats in order to stay alive.

Why do farmers engage in such prodigious waste—using protein to create protein? Fishmeal producers claim that their feed provides livestock with "essential nutrients" not present in grain. Yet when fishmeal stocks declined after the catastrophic Peruvian anchovy collapse, and fishmeal became expensive, farmers readily turned to soybean meal to make up the difference. Herein lies the more believable reason why fishmeal periodically gains popularity on the farm: Farmers feed fish to their pigs and chickens when fish is cheap.

The flip side of the question—whether people can eat the fish we feed to animals—does not lend itself so easily to answers. Some of the fish caught for animal feed are edible by humans, and in fact a few countries have begun to divert their catch of these fish into canned goods for human consumption. Since the late sixties, fishing agencies and food-aid organizations have also experimented with a ground-up, dried fish flour—"fish protein concentrate"—that could be made from otherwise inedible fishmeal and could fortify bread, cereals, and milk in the Third

World. Mired in politics, delayed by the costliness of the necessary processing, the idea of fish protein concentrate was finally defeated by the simple fact that even starving people wouldn't eat it. "Telling people that food is good for them and asking them to forget the taste does not seem to work," one agency executive lamented.

Fishmeal thus presents a moral quandary. While it is easy to understand why Third World people refuse to be fed like pigs, it offends anyone's moral sensibilities to know that while the Peruvian anchoveta fishery was producing the world's single greatest supply of fish, almost all of which was turned to chicken feed, children in Peru suffered and died of protein deficiency. Yet even these moral predicaments should not obscure the central fact: An industry that turns one third of its produce over to farmyard animals is not devoted to feeding people.

If we really wanted to feed the hungry world with fish, then why have industrialized nations depleted the waters of the hungry world? Two short sentences recently uttered by a U.N. Food and Agricultural Organization official speak volumes about our real intent. "Six million tons of fish were caught in the marine fisheries of the [African] continent in 1978," he remarked in a speech urging intensified fishing effort. "I would mention in passing that only 1.8 million of that total was caught by African fishermen."

"I should mention in passing": These five words encapsulate what has happened during the past three decades of fishing—with industrialized nations treating the underdeveloped world as an afterthought. In the sixties, just at the time wealthy nations began to find nearby northern waters empty, they looked to southern seas and saw that the richest fishing grounds left on Earth were harvested only by local artisans, without factory ships, without refrigerated storage, without processing facilities, without even feeder roads to take fish to interior populations. The industrialized nations officially offered to "help." Soon an international fleet of purse seiners and trawlers descended on the upwellings off northwestern Africa, Namibia, and South Africa. The USSR, Spain, and Japan between them besieged schools of horse mackerel, squid, hake, sardinella, and pilchard, streaming their take out of African waters and back to home markets.

Next, fishing officials urged industrialized nations to deploy the newly invented "automatic homing gear" on their supertrawlers in the Indian Ocean—to aid, of course, "human populations sorely in need of protein." By 1969, the people of Kenya—"sorely in need of protein"— had managed to land only seven thousand metric tons of fish from the Indian Ocean, while the Soviets harvested three times that much, and the Japanese, fifteen times. From 1975 to 1977, the literally precious catch taken from northwest African waters, excluding tuna, brought in an estimated jackpot of nearly $1 billion—three quarters of which was pocketed by advanced nations.

During the very years the FAO incited industrial trawlers to converge on the Indian Ocean with their "automatic homing gear," *Calypso* sailed the region. We came upon a group of Maldivians gathering food from the sea with their elegant handmade sailboats, as their ancestors had done generations before. Inside their crafts they had fashioned live-bait compartments—by drilling holes below the waterline! The practice did keep their bait alive, but it didn't do much to keep their boats afloat. They had styled an ingenious method of fishing: The fishermen perpetually bailed. As they tossed their buckets of water, they disturbed the sea surface. Combined with what observers call "paddle flicking," the splashes mimicked a feeding frenzy of little fish, attracting larger, tunalike bonitos. The islanders' shiny barbless hooks, flashing in the water like silvery fish fry, caught the bonito in one last deception. Some of these islanders dutifully recited for us their ancient law: "Take no more from the sea in one day than there are people in your village. If you observe this rule, the bonito will run well again tomorrow."

It is these forgotten fishermen who must compete with the five-thousand-ton trawlers of the industrialized world, not just in catch but also in quality, processing, and price. There remains little to compete for. In the sixties the USSR, Spain, and Japan besieged the stock of Namibian pilchard; by 1978 the pilchard stock had collapsed. The once abundant hake stock that had attracted the Soviet, Japanese, and South African fleets was also overexploited. After overfishing, the southwest African sardine catch abruptly tumbled by half.

As if it were not enough to ignore the Third World's people while fishing out their waters, the fishing tycoons up on their towering bridges remember the artisanal fishermen all too well when the waters are suddenly empty. Unbelievably, it is to the artisans—to Moses and the Maldivian paddle flickers—that the fleet operators then accusingly point. Even though the Soviets and Eastern Europeans used large-scale fleets of trawlers to fish sardinella off the shores of Africa, when the inshore catch collapsed, artisans were blamed for having caught the juveniles that remained. These local fishermen, according to expert opinion—with no mention of the offshore squadrons—"now need to be regulated."

After thirty years of "help" from the industrialized world, millions of small-scale fisherfolk, the United Nations now reports, presently face the "danger of being increasingly marginalized and impoverished." They still lack refrigeration, still lack storage facilities, still lack roads, still lack skills. Yet advanced nations have left them with something: They have left them with no fish.

The depleted Third World waters, the fish sold for swine feed, the squandering of unprofitable fish—all point to the fact that we deplete world waters not in quest of food, but in quest of profits.

Yet fishing does not now, nor has it ever, undergirded the economy. In most countries, fishing represents less than 1 percent of the gross domestic product. French fishermen contribute just one tenth of 1 percent to the GDP; farmers bring in thirty-nine times as much. In the United States, fishing infuses the economy with just eight hundredths of 1 percent of the GDP; farms provide ten times as many jobs.

Even more ominous for the economy: A fisherman who collects an income today cannot count on it tomorrow. As fishing grounds collapse, local economies jerry-built upon them give way as well, taking jobs, factories, and all the ancillary businesses with them. Once economic king of the North Sea, the British drifter fleet and all the jobs it offered crews ceased to exist when the herring disappeared. When cod off Iceland languished, so did thousands of fishing jobs in far-off Yorkshire, England. The waning Pacific salmon runs have thrown hundreds of fishermen in British Columbia, Canada, into financial ruin as well. As we filmed a documentary on Canada, we saw the debris of overfishing strewn even across the shores of the Great Lakes; abandoned fishing villages testified to a fool's rush for trout just as hauntingly as ghost towns in the American West tell of a fool's rush for gold.

Nor have developing nations evaded overfishing's economic repercussions. When Peruvian fishermen eradicated their anchovy industry, their nation lost some $340 million, nearly a third of its annual foreign exchange earnings.

Yet the immediate costs of a collapse—in unemployment, job retraining, and relocation—are not the only toll the fishing industry takes on the economy. The startling extent to which fisheries drain a nation's economic bloodflow is evident in one simple fact: The majority of the world's fishing industries, though barely raising an operating profit, persist only with some form of government subsidy.

West Germany would have had to sell its modern but financially failing fleet to South Korea and China were it not for the subsidies its taxpayers doled out in 1985. American taxpayers bear the additional budgetary burdens contributed by the zero-interest tax deferments and interest-free loans their government offers to fishermen. Because the depleted North Sea herring fishery can no longer support Norway's vast fleet of purse seiners, Norwegian taxpayers now guarantee a minimum weekly wage to their fishermen. In order to secure twenty thousand processing-factory jobs threatened by overfishing of their waters, Canadian taxpayers picked up a $425 million tab to buy and reorganize the companies—pouring back into the fishery about one third of its annual export earnings. And still the companies sink in a quagmire of serious financial difficulties.

The most irrational subsidies are surely those that taxpayers dole out for advertising campaigns to increase fish consumption in spite of depleted waters. Although advertising for fish consumption is like advertising for gas guzzling, Canada, with its bankrupt fish factories, subsidized an "Age of the Fishery" seafood promotion campaign. Norway—after the lowest catch in twenty years from its severely overfished waters—assisted private companies in a plan to boost fish consumption. Japan, which insists that it must harvest its overfished waters to supply its people's protein needs, poured tax money into a promotional pamphlet publicizing cooking methods to lure more diners to a table its own seas even now cannot supply.

The industry pulls fewer fish out of the water as taxpayers throw more money into it. The consumer pays for the same fish over and over: first in price supports; then in the higher price itself; then in grants and loans and advertising to tempt fishermen to deplete waters further; and finally in the even higher prices commanded by ever-more-rare fish. No one could deny sympathy to the fisherman faced with loss of his boat, his home, his livelihood. Yet subsidies do not preclude the costs of unemployment; they postpone them. They even induce them: Depleted waters that cannot support a fishing industry today will surely not support it after they have been further depleted, thanks to subsidies, a few years hence.

Official attempts to staunch this flow of dollars—by limiting fishing effort through quotas—have failed. Why? Confronted with quotas, subsidized fishermen simply buy more boats. Quotas become a finish line; the fishing season, a race in which fleet owners speed toward their fair share, at least, or the biggest share, at best, of the total allowable catch. The owner with an overexpanded fleet gets an oversized share of fish in the shortest time. When authorities then license ships—limiting the number of boats that can enter a fishery—owners engage in an even more cutthroat competition: They buy higher technology, jamming $40,000 of electronics into $10,000 boats. Any "winner" finds his victory only evanescent, as others soon buy more boats and more equipment and simply overtake him. Together the competitors achieve each region's total quota earlier and earlier. Officials have closed some grounds after a fishing season of only a few hours. Fishermen then remain idle, nervously eyeing their marina of mortgaged boats. Processing factories ashore, which have multiplied to accommodate the massive tonnage of catch that comes at the beginning of the season, have nothing to process by the end of it; nonetheless they must maintain their machinery. The more money everyone loses, the more equipment everyone buys. The laws of supply and demand—according to which the knockout costs of increased fishing effort would throw some owners out of the ring—do not apply in an industry in which tax subsidies encourage more investments. The industry rides trapped on a nightmare carousel, dragged over and over around the same cycle of mistakes.

Their boats burdened with elaborate sonar equipment, fishermen off west Scotland in 1980 regularly took in their whole week's quota in just one night, leaving six days ahead with nothing to do, nothing to catch, nothing to earn. In the Gulf of Alaska, fishermen had to spend ninety-six days, in the mid-seventies, in order to catch fourteen million pounds of halibut; by the early eighties, the fishery became so jammed with competitors that the overexpanded armada landed the same catch—in spite of a fish decline—in just sixteen days. A Yale scientist, quoting an estimate of $4 to $5 billion in unnecessary expenditures on fishing in the North Atlantic, remarked that this ocean—both in the east and in the west—"has been and continues to be fished with far more effort and far fewer results than make any sense."

This wasted money in the fishery translates into shocking waste for society. While the fishery's contribution to the gross domestic product is small, its net national product—real profits, minus costs—is infinitesimal. In the fifties, U.S. fishermen already netted less than half as much as average workers in other industries. Throughout the following two decades, the average American worker's net gain more than doubled, while the American fisherman's net income rose by only 50 percent. Worse, though these were the very decades during which fishermen poured investments into revolutionizing their fleets with technology, every cent of their measly 50 percent net gain was due not to an increase in per capita catch—which actually fell—but to inflation. The new fishing technology produced absolutely no returns whatsoever for American society, though it probably did wonders for the Japanese electronics industry.

If rampant fishing drains the economy—and fails to save the starving—then what urgent need drives fishermen to exhaust fish stocks, compromising their own renewable resource and sabotaging long-term economic gain for society?

The industrial fisherman does not fish "for our society." He fishes for himself. If one hundred fishermen each own one boat, and one of them then buys five more, ninety-nine individual fishermen's profits drop; but the boat buyer's profits immediately increase. Everyone's loss is one man's gain. The only profitable attitude is opportunism. Each fisherman refuses to conserve for others out of the certainty they won't conserve for him. No economic theorist could more succinctly describe the motivation behind fishery economics than one owner of a fishing boat on the quay in Yarmouth. Biologist David Cushing asked the old salt what he would do if he were given unlimited power to save his business as the ancient East Anglican herring fishery was collapsing. The fisherman growled, "Burn every bugger's boat but mine."

So also is it a delusion to believe that the industrial fisherman fishes for long-term economic gain. He fishes for cash on the barrelhead. The fishing industry, in effect, simply weighs the profit potential of fish caught and sold—money invested and earning interest today—against the value of forgoing immediate returns so that some fish can reproduce for tomorrow. The more slowly a fish species bears offspring, the more lucrative it is to take the money and run.

Obviously this disrupted fishery sorely requires leadership, some thoughtful authority that could end the anarchy. Yet to whom have we ceded local control of fisheries? To the fishermen, to the processing-factory owners, to the boat builders, even to individual investors who have never gone near a dock but have been lured into the financial whirlpool by the siren song of profits. Again and again, rational observers puzzle: Why does an industry that contributes so little to society wield such disproportionate influence? The answer is simple: The strength of the fishing lobby derives from the enormity of its potential losses. Fishermen wield a great deal of influence because they have a great deal at stake. Harassed local politicians do not think in terms of future fishing seasons; they think of coming elections. Fish don't vote; fishermen do. No matter how great the long-term gains of regulation, it is the fishermen's outcry over short-term hardship that politicians and their local regulators hear. While taxpayers may unwittingly share in the burden of the current system, they are unlikely to know or care enough about its pitfalls to demand a change.

The chaos that characterizes local regulation characterizes global regulation as well. Some thirty-nine different international agencies, whose responsibilities leave gaps as often as they overlap, preside over the world's one fishery, in the world's one ocean. The actual intent of founding nations is evident in the enforcement authority they endow on the agencies: none whatsoever. The agencies can generate at best only nonbinding "gentleman's agreements." Staff biologists and administrators know they must tread lightly or the member nations will break the voluntary accords and simply quit. James Joseph, director of investigations for the Inter-American Tropical Tuna Commission—which has suffered just such defections—has undauntedly scanned the hordes of agencies dotting the regulatory horizon and has concluded outright that they fail to solve the problems of fish depletion through "lack of coordinate initiative, lack of sufficient authority," and most of all, "the lack of political will to get the job done."

The history of these agencies has thus been written in long and confusing lists of successive regulations, bound by one unifying trait: None of them has worked. To each regulatory failure, the industry tags another rationale. Setting quotas and insisting on licensing for ships have failed, as we have seen, because they lead to overexpansion of fleets. Rules ordering larger mesh in trawl nets, to protect the juvenile fish of diminished species, have also failed; fleet operators contend that big mesh prevents them from catching the perfectly abundant adults of tinier species. Rules requesting tuna purse seiners to use *smaller* mesh— to save the dolphins that swim above tuna schools, become ensnared in nets, and drown by the thousands—have likewise failed among most nations: Fishermen argue that the prescribed nets are heavier, more expensive, and bring aboard no greater catch of tuna for the extra cost.

The real explanations why rational rules never seem to work are less complicated. First, "regulatory" agencies do not regulate; they only arbitrate. Member nations take so long to negotiate the most beneficial terms for their own fishermen that the fish populations themselves— overharvested through all the years of bargaining—meanwhile collapse; agency biologists are forced to spend their time and expertise, in the words of one scientist, not in averting disaster but in conducting "sophisticated postmortems."

Second, when politicians finally act, they set quotas that are not quotas, but rather permits to deplete. Marine biologists base their recommendations for quotas on the scientific fact that a certain amount of

harvesting actually stimulates fish populations to grow faster and reproduce more often. They can determine the "maximum sustainable yield,"
the absolute limit of catch fishermen can safely harvest before under-
mining nature's compensation and leaving too few fish behind to reproduce. Nature, of course, does not always behave so neatly, and scientists
can be wrong, but more often quotas fail because nations belonging to
fishing agencies base the allowable catch not on biological recommendations but on political advantage. Their method has been described with
the euphemism "passive management"—they set quotas so huge that
the myriad of fishermen already exploiting a fishing ground will all be
accommodated.

This melodrama of member nations' fiddling, while the fishery
burns, was most vividly enacted on the herring grounds in the North
Sea. A. C. Burd, a biologist who submitted findings to pertinent fishing
agencies, has recounted the organizations' interminable series of nationalistic delaying tactics. In 1956, the International Commission for the
Exploration of the Sea (ICES) first held a scientific meeting to discuss
the declining herring catch. Political appointees could reach no agreement but decided to undertake a study. Four years later, the ICES set up
a "working group" to determine the effects of fishing; "this was not
achieved," as Burd put it, and a second group was set up to review the
problem. Another four years later, the second group stated in a report:
"Though there was a considerable measure of agreement on the evidence, it was not considered sufficient to indicate what action [to] take."
One year later, ICES members agreed that only one stock of herring
would benefit from reduced fishing—but that the "cause of the failure"
of another stock had been induced by nature. Two years later the ICES
discussed closing parts of the North Sea to "all herring fishing," but
"some delegations required more scientific evidence . . . others stressed
that economic considerations should be given weight."

During these years, the catch of the overpressured herring stock fell
98 percent from its all-time high. Burd concluded: "A major fishery was
lost while the search for absolute certainty of the cause of the decline
was sought by management." The ICES member nations—so eager to
protect their industry—had in fact succeeded in destroying it.

The third unspoken reason why regulations don't work: While politicians render rules meaningless, the fishing industry itself makes a mockery of them, doing everything within its power to give scientists fraudulent information on which to base their recommendations. Fishermen undermine even basic statistics—falsifying catch reports, grossly underrating their take, rarely if ever recording the fish they dump overboard. In order to keep inshore grounds, where fish spawn, open to fishing as long as possible, California fishermen are suspected of having falsely reported that most of their anchovy catch came from distant waters. Many fishermen refuse even to report their catch at all.

The industry has refined evasion into an art form in the markets as well as on the open waters. Not long after the war, I took the Mistral train from Paris to Marseilles at five o'clock one morning. Wearing a dark suit and carrying an attaché case, I arrived several hours before my first appointment; to kill time, I wandered into the fish market in the Old Harbour. With its bustling activity and incessant haggling, the place sounded like a cage of screeching parrots. Suddenly, a corpulent woman who had anchored herself at the entrance saw me. In an instant, she put her hands to her mouth and imitated a siren. The market abruptly fell silent. Fishmongers lit cigarettes, shifted their weight foot to foot, placed their hands on their hips, gazed up at the ceiling. I kept walking, a little embarrassed, amid the stony silence. I made a full circuit of the market—among fishermen so still they seemed frozen in time—until once again I reached the forbidding figure of the fleshy woman, still ensconced at the entrance, and passed her on my way out. Immediately, she put her hands to her mouth and bellowed, "End of the alert!" The full roiling noise of the market bubbled up again. I understood: That imposing dame, who I'd guessed was certainly the market sentinel, had mistaken me in my business suit for a price controller.

Just as much creativity is used to evade regulations at sea. In 1985, I commuted often between Paris and the shipyards in La Rochelle, France, where I was building the windship *Alcyone*. One day during the commute my airplane seat companion was an attorney representing a Spanish vessel. The boat had been in the headlines at the time, having been seized by French authorities for fishing illegally in French waters.

The lawyer shared with me his firsthand account of the entire misadventure, outlining the standard Spanish modus operandi: While dozens of Spanish boats descend on a forbidden fishing ground, he said, their crews keep an eagle eye for the approach of the French Navy. As soon as a police vessel appears in the distance, all boats but one scramble for the high seas. The fishermen aboard the remaining boat quickly throw their nets into their propellers; when the patrol boat arrives, the Spaniards plead innocence at having trespassed, explaining that they were only out of control and drifting. French authorities confiscate the nets and catch and levy a fine. Meanwhile, the hundreds of Spanish fishermen waiting just over the horizon divvy up the fine as an operating expense, each paying a negligible amount to compensate their dupe for his loss. The Spanish government's response? It now officially reports, in the gracious patois of diplomacy, that it will grant its fishermen a special subsidy for "expenses arising from compliance with conditions imposed by foreign governments for continuing to admit the Spanish fleet to their ground."

As catch and income decreased through the sixties and seventies, relations between fishermen of different nations became rancorous. The contentions between the fishing industries of Spain and France finally drew blood when Spanish trawlers trespassed into French waters and France sent warships that used guns to drive them away. Canadians and Americans together railed against the "floating international community" that had pillaged their Atlantic haddock and flounder industries. U.S. fishermen accused Soviets of steaming through fleets of small American boats and destroying gear. Chile, Ecuador, and Peru unilaterally claimed territorial boundaries to prohibit foreigners from catching tuna that migrate off their shores; they seized American tuna boats that defiantly crossed the lines, jailing skippers, pelting crews with garbage, even firing on the boats. On the other edge of the world, Iceland and Britain battled in a bitter "cod war." When Iceland tried to ban foreigners from fishing two hundred miles off its coast, to protect the nearly eradicated cod stocks of those waters, British fishermen promptly arrived in the forbidden area. Accusations later flew in international court: that U.K. naval units, attempting to enable British fishermen to continue trawling, had rammed Iceland's coast guard vessel; that the Icelanders had aimed one of their ship's guns at the British bridge and had fired other weapons repeatedly at close range. The United States nervously tried to intercede, fearing that Iceland—site of a crucial military intercontinental ballistic missile alert base—would storm out of NATO in a fit of rage over the fishing imbroglio. Now the fishing industry was threatening even the international peace.

In stunning contrast to the sordid disorders of the fishing world, Arvid Pardo, Maltese delegate to the United Nations, startled the global community with a noble proposal. In 1967, he urged the United Nations to declare the sea and its resources "the common heritage of mankind," not to be appropriated by any one nation for its own exploitation, but rather to be safeguarded by every nation for the benefit of all. Inspired by his words, members of the U.N. General Assembly unanimously approved Pardo's declaration and instituted a Seabed Committee to propose laws that would embody his magnanimous concept. Alas, if ideals can easily impassion an audience, a committee can readily deaden the ideals as members attempt to agree on practical ways in which to implement them. Pardo's vision was doomed from the start: It was too generous. I and others who cared for the sea ached as we watched his proposal be downgraded, step by step. Economic considerations and the political imperatives of individual nations slowly corrupted his dream. Nations finally responded to the call for one international authority by proposing that each nation wield its own authority. By the time the first substantial session of the Law of the Sea Conference convened, seventy-six countries had, in anticipation, already rushed to claim they "owned" their coastal seas. The conference's 1975 negotiating text officially recognized this idea of ocean ownership; it proposed the "Exclusive Economic Zone" (EEZ), a region including all waters and the seabed two hundred miles from each coastal nation's shores, in which the country is free to allow its fishermen any fish harvest it wants; it may also sell or barter additional fishing rights to foreign vessels. Today such proprietary claims by coastal nations enclose the waters that generate 99 percent of the world's total catch.

Fishing lobbies exulted at the idea of the EEZ. With foreign nations held back from first grab of the domestic catch, local fishing industries felt that economic nirvana had arrived. For its part, the U.N.'s Food and Agricultural Organization predicted that this new idea for overseeing the fishing industry would prove better than all the old ones, that the two-hundred-mile EEZs would give us "a window in time in which to get things right."

Many years have passed since nations first began to claim their individual ownership zones. Not much has turned out right. The new regime has not alleviated but has aggravated overfishing as well as over-expansion. National fleets have regarded the Law of the Sea not as a mandate to protect their resources but rather as an open invitation to plunder them. Finally rid of foreign competition, they have interpreted the Law of the Sea as the call of the wild.

The proof lies in the burgeoning fleets. In 1977, in the wake of two-hundred-mile claims staked on continental shelves, I had a foretaste of things to come. With *Calypso* working off Tunisia, I invited the country's minister of fisheries to come aboard. He eagerly told me of the way he was sure he would double his catch: by doubling his fleets. The minister was not alone in his mistake.

The fisheries of other nations have greeted the ownership of their national waters with the same misguided zeal. The USSR and Japan—the two nations that already possessed the largest fishing fleets—both immediately built an exceptional number of new ships. The Netherlands and Italy swelled their ranks of fishermen and subsidized more boat building, even though stocks of their targeted species were steadily sliding downward. Sweden also paid for more vessels as its catch diminished.

Furthermore, the new EEZs have not protected Third World waters from industrialized nations but have intensified the temptations to deplete them. The United Nations fantasized that the new two-hundred-mile limit would bestow the developing world with fisheries "worth some $1.9 billion a year." Immediately, industrialized nations materialized at Third World doors, promising to "provide coastal states which do not possess the capacity to harvest the total allowable catch with the possibility to profit." As in the past, these promises may turn out to be empty.

This so-called help has taken the form of "joint ventures," a deal in which one nation can gain fishing access to another's waters by paying a fee, agreeing to a political trade-off, or employing the people of the coastal country, teaching them to fish and offering technical assistance. Industrialized nations have found the arrangement more than satisfying: Spain, often unwelcome in the seas of advanced nations due to its proclivity for ignoring quotas, had 180 large vessels conducting joint ventures in 1982 alone, with the government subsidizing a fleet expansion so that its fishermen could pursue the already heavily exploited yellowfin tuna in African waters. France, which subsidized a new fleet of purse seiners complete with heliports, and the United States also aim to aggrandize their catch of the overexploited tuna species off Africa. The African continent's hake stocks, meanwhile, after a dramatic decline in 1982, may recover just in time for Russia, South Africa, and Spain to exploit the fish in Namibian waters.

The Third World nations themselves, once eager to follow the EEZ rainbow, are beginning to discover that the promised pot of gold may be mostly mythic. Mauritania collected some $29 million in fees for joint ventures during 1978 alone, but the fees fell far short of the fishery's potential worth—theoretically several hundred million dollars—had the country been both technologically and scientifically equipped to manage the resource wisely by itself. Yet when advanced nations have offered assistance and training, they have assisted the poor countries by giving them more boats and training them to commit all the sins of overcapacity that exhausted so many northern fisheries. The "aid" projects have supplied new motors, for example, to natives all along the West African coast; the artisanal fishermen, with their souped-up canoes, have increased their catch for a while, but not for long, as stocks soon succumb to overpressure. Several such "aid" projects, according to one review, "have resulted either in a further impoverishment of fishermen or a reduction in employment."

Finally, the new regime has not soothed but only incited international enmity. Because the EEZ arrangement allows each nation to allocate foreign fishing rights for whatever reason they choose, regardless of biological considerations, fish have become a mere bargaining chip in the international poker game. New Zealand and the United States cut their fishing allocations to the Soviet Union, not for the good of the fish, but in retaliation for the invasion of Afghanistan. European and American strategists opposed Norway's fishing pact with the USSR because they feared the concentration of Soviet military strength in the Barents Sea. An American senator, angry because Japan rapaciously harvests fish the U.S. industry considers "American," proclaimed the Japanese to be the AIDS of the high seas. The Irish Navy fired on a Spanish trawler catching "Irish" fish illegally; the trawler sank and sixteen Spanish crewmen had to be hauled out of the drink. Meanwhile, the passions that ignited the two-hundred-mile movement in the first place continue unabated, with Chile, Ecuador, and Peru bickering with the United States over who "owns" the tuna that swim hundreds of thousands of miles back and forth over oceans, oblivious of the boundaries politicians are trying to impose upon them.

Nearly a decade after most nations claimed their two-hundred-mile zones, announcing that they would thus lend order to the world fishery, we still have anarchy on the oceans. Only now we have egotistical anarchy: an anarchy of 135 individual coastal nations each thinking solely of itself, each promulgating its own law, each ensnared in lawlessness of its own making.

Today fishing organizations scurry to identify new scapegoats, to make new proposals. Now they say the idea of "maximum sustainable yield," basing quotas on fishes' reproductive capacity, is to blame. The new cure-all that bureaucrats propose: "optimum sustainable yield"— quotas set in consideration of "the economic needs of coastal fishing communities," as though "economic needs" have not been the sole consideration before. Officials also blame the two-hundred-mile limit itself. Prior to the coastal zones, say fishery spokesmen, the entire world ocean was debased as "common property"; international fleets treated the seas as vandals do a public park, destroying resources for personal benefit knowing that the costs would be borne by others. But now, complain the experts, the two-hundred-mile zone itself has become just a smaller version of the public park, vandalized by fishermen of the "owner" nation just as it once was by international flotilla. The new cure-all that officials propose: assigning even *more* detailed property rights, giving quotas not just for each fishing ground but now even to each single fishing boat within it. With these "vessel-allocated quotas," they are sure they can remedy the free-for-all that sociologist Garrett Hardin dubbed—as agency officials eagerly echo—the "Tragedy of the Commons."

There is indeed a tragedy in our fisheries, but it is not a "Tragedy of the Commons." The tragedy lies in the fact that we have steadfastly refused to respect the fisheries *as* a commons. While the oceans once were *nobody*'s land, instead of making them *everybody*'s land, we made them *somebody*'s land. Basing future regulations on the "economic needs of fishing communities" is not a new solution; it is the old problem. Nor are vessel-allocated quotas likely to work. Quotas failed in the past because we never had quotas; unenforced quotas don't exist.

We will safeguard our fishery only if we stop disobeying all the political laws and start honoring a few natural laws—if we simply cease our obsession with rules intended to protect the fishermen and begin respecting rules that also protect the fish.

We should rotate trawling, as farmers rotate crops. Because the massive trawl nets scrape the bottom, they destroy the fauna where fish spawn. If fishermen trawled in an area one year, then allowed it to lie fallow for several years, the fauna could revive.

We should reduce the use of purse seines. Lines at least take only individual fish from a school, while purse seines wipe out a school altogether, leaving no stragglers to reproduce.

We must never catch fish where and when they gather to spawn. By persistently and systematically ravaging stocks as they reproduce—simply because it is easy to catch them in the middle of their nuptial ceremonies—fishermen cut off the very branch on which their whole industry is perched. When we took *Calypso* to the southeast Caribbean, we came upon some sardine fishermen. The sardine, they said, had played a curious trick. The fish had once thronged the Bay of Cumaná in such multitudes that investors had built a large cannery onshore; then, after a few years, the fishermen told me quizzically, the sardines had for some reason "left." The disappearance didn't seem so mystical to me; if a fisherman hauls in nets of females filled with roe, he must know he is interfering with procreation and thus is precluding perpetuation of the stock. The fishermen could not admit they had squandered their own assets. It was the fish! The sardines, they said, had picked a new location for their nursery, and now anyone who wanted to harvest tons of sardine had to go to the waters off the island of Margarita. In fact, the fishermen had laid waste to one spawning ground and then, off Margarita, were laying waste to another.

In order to protect the resources we must also begin to respect the intricate dependencies fish have upon the other creatures in their communities. When fishermen ignore the complexities of interaction, and overharvest one stock, other stocks—and sometimes the entire pyramid of life—come crashing down as well.

In 1979, as *Calypso* tracked humpback whales migrating northbound past Newfoundland, we saw how whales became literally entangled in the woes of the capelin fishery. The little capelin had once thrived in such bounty off Canada's Grand Banks that it had served as food for all the major commercial fish—cod, Atlantic salmon, Greenland halibut, haddock, and several kinds of flounder. Hungry humpback whales, on their arduous journey from tropical birthing grounds to Arctic feeding grounds, sustained themselves on capelin too. When the vast international fleets off Canada's shores descended on this fish, capelin stocks collapsed in just six years.

Instinctively, starved whales veered off their usual track, to sate their hunger on the capelin stocks that still survived inshore. Unfortunately, inshore fishermen were harvesting cod there at the same time; the cod fishermen had set more gear in Canada's inshore water than ever before. As we traveled past Newfoundland, we dodged and darted with *Calypso* between mile-long walls of cod nets, set by the hundreds. We maneuvered through; not every whale in pursuit of its capelin meal was as fortunate. All that summer, humpbacks had been drowned by nets or injured in collisions with fishing equipment. Enraged at the thousands of dollars lost in ruined fishing gear and interrupted fishing each time a whale became ensnared, the fishermen we met joined in an angry refrain: "You feel like killin' 'em, you do." "The first thing sometimes, ya wanna shoot the damn whale." Sometimes, we heard, they did.

Such domino destructions point to the most obvious truth: If we are to safeguard our fishing resource, we must stop our inexorable progression in depleting species after species, to ground after ground, always believing some new fish stock will magically appear around the next corner, in the next bay.

No fiasco better maps our advance to nowhere than the overall whaling saga. After scouring the northern seas, depleting the slow-swimming gray whale, the right whale, and the bowhead, whalers turned to the mighty blue whale, the biggest creature ever on Earth. Next whaling ships focused on the Antarctic, where they severely depleted blues and then trained their guns on fin whales. When Antarctic fin whales foundered, the International Whaling Commission—another agency beleaguered by politics and bereft of authority—urged whalers to chase sei and humpbacks. On, then, to the northern Pacific, where Soviets increased their high-seas whaling expeditions fourfold, and where blue and humpback populations ebbed. Next: the sperm whale. In a failing market, the British, Norwegians, and Dutch unloaded their whaling fleets on the Japanese, and the process of amortization and exploitation began all over again. By 1973, when *Calypso* sailed to the Antarctic, we used our helicopter extensively to seek out any whale to film. We operated in a zone the International Whaling Commission had that year declared a preserve for the giant mammals; if there were any whales to be seen, we would see them there. Over Earth's most fertile grazing grounds for whales, we logged more than three hundred hours in the air, over two and a half months, and saw only two humpbacks and perhaps two dozen of the little seis. We saw not a single fin. The only great blue whale we saw was the one we built, from giant, bleached bones discarded over the shores by whalers who considered blubber and flesh as commodities but the rest of a whale just so much debris. We became amateur paleontologists, piecing bones together on the ground; the giant rostrum took all hands to put in place. The final skeleton measured more than ninety feet in length. As I gazed over the interminable spinal sequence, I thought about how the industry that began by expunging this colossus is now, with nowhere else to turn, mustering for the final assault—against the three-inch krill.

On our helicopter surveys, we saw red blotches on the surface of the sea, a sign of swarming millions of these carrot-colored, shrimplike creatures. The primary food source for baleen whales, krill populations—free of their major predators since whale stocks were butchered—have been exploding.

And so the mathematicians of industry have gone to work, making computations. They have calculated that 1.1 million baleen whales—45 million tons of biomass—once ate 190 million tons of krill; by 1973 the surviving whales, with a biomass of only 9 million tons, probably ate no more than 43 million tons of krill. Ergo, a huge annual surplus of at least 150 million tons of krill must remain—a weight equal to twice the present catch of all fish around the world!

Yet even as ships forge through the Antarctic for a krill harvest, no one has the slightest idea if a krill "surplus" really even exists. No one knows what consequences a massive krill hunt will have on the narrow and fragile pyramid of life the Antarctic has built upon its krill foundation. What future for the sperm whales and birds and seals that feed on squid, which feed on krill? What future for the crabeater seals, the Adelaide penguins, the fish, the southern fur seals, all once overexploited but now, through whaling's backhanded blessing of more krill, are each slowly multiplying? What future for the oceans over the entire earth, nourished on the nutrients released as krill die, sink toward the bottom, and sweep along with the deep currents that wash around the globe?

While we were diving in the Antarctic, dazzled by the variety of strange creatures we filmed above and below the surface of the sea, I thought that perhaps the last continent explored by man could be the first continent he would not plunder. I was wrong. Already the trawlers have arrived. Krill catches burst from 3,300 tons taken in 1976 to 530,000 tons in 1982—most of which were landed by Japan and the Soviet Union, the two last major whaling nations, which, having reduced the whales to a nonprofitable level, could now preclude the creatures' return by harvesting their food. Fishery spokesmen wax ecstatic: With krill we can triple our catch; with krill we can feed the hungry. Good luck to them. In the Antarctic we caught fresh krill, fried them, ate them, and instantly suffered diarrhea.

Modern fishing has depleted the streams and then the bays, the gulfs, and finally the seas. From one sea, fishermen have moved to another, until today they have painted themselves into the last corner, exhausting the ultimate fishing ground, the global ocean. And still we say it doesn't matter, that the limitless sea holds untold new species, that the last untouched stocks of Earth—swimming under the frozen Antarctic—will somehow be able to sustain the imminent industrial blow without collapsing as other stocks have collapsed before them.

In order to profit in the present, fisheries have forfeited the future. But the future is now; the losses are ours. They need not be. If we respect the biological laws of the fishery, after a few years we might even double our harvest, actually achieving those catch projections that now are only chimerical. The mechanism to implement these measures already exists; we need only alter our perspective. We can turn our famous EEZs, the national zones of exploitation, into NZRs—national zones of responsibility. We can stop insisting the resource belongs to all nations to devastate, and begin realizing it belongs to all nations to protect.

While this solution is theoretically within our reach, the question remains whether we have the will to grasp it. We have little chance if we behave in the future as we have behaved in the past. I felt my own optimism waning shortly after Venezuela's ambassador to the United Nations invited me to speak at the opening of the Law of the Sea Conference in Caracas. The ambassador asked me to outline my expected remarks. By that time, the idea of declaring the sea to be the common heritage of humankind had already begun to erode, with each nation claiming a piece of the ocean as its own. "No one can 'own' water that constantly moves from one shore to another," I told the ambassador. "I will say that in order to put an end to the fishing anarchy, we must create a World Ocean Authority, which would establish rational rules that each coastal nation—taking responsibility for its own two-hundred-mile zone—would then enforce."

Two weeks later the ambassador called back, infinitely embarrassed. He had relayed my plans to conference officials, and they had asked him to withdraw my invitation to speak.

I went my way. The officials went theirs. And not too far away, across some fishless waters, Moses went to bed, again, having eaten almost nothing.

CHAPTER EIGHT

SCIENCE AND
HUMAN VALUES

I always wondered what lies under the sea—not under the surface of the sea, *in* the water, but *under* the water, under the sands of the seafloor. Sonar's cryptic messages concealed these secrets; its taunting signal only bounced back off the bottom, simply profiling the topography of undersea terrain. My longtime friend, the electrical engineering genius Dr. Harold Edgerton of MIT, shared my intrigue with the unknown geological provinces buried below. He and I spent hours lowering cameras into the depths, but even film of the deep-sea floor didn't quell our nagging need to know what mysterious strata the bottom muds engulfed. At last Doc Edgerton devised a way to peer into these subterranean recesses: He invented instruments he called *boomers* and *sparkers*, which could hurtle powerful signals more than a mile down into the bottom. The intricate electronic echoes that reverberated back would reveal the various marbled layers of sediment that blanketed our planet's sunken crust. I eagerly outfitted *Calypso* with the new equipment, hired a crew of electronic specialists, and set out to find what tiers of earth undergirded the Mediterranean basin.

As in most oceans, steep underwater canyons at the Mediterranean's coastline dropped precipitously to an abyssal plain—a vast expanse that usually stretches out as smooth and level as a ballroom floor. But then, halfway between Nice and Corsica, we noticed a few low, barely discernible mounds swelling up slightly from the table-flat bottom. They looked like nothing more than ordinary drifts of sand. But were they? Sheer curiosity took hold; we stopped; we crisscrossed the mounds from above, on *Calypso*; we bombarded them from every angle with our signals. The sparker readouts were like X-rays of planet Earth; they showed that deep within the bottom sediments lay a vast, horizontal layer of hard material, spreading out for miles in all directions. But just below us, this layer had buckled, rising in fingerlike columns that pushed upward, their tips nearly touching the surface of the seafloor, forcing up our mounds. Salt domes!

Geologists had long believed that the Mediterranean was once an inland sea; here was physical evidence. Our discovery indicated that, without its present inflow from the Atlantic, the Mediterranean had at one time actually dried away; the evaporated seawater had left behind a blanket of salt perhaps 245 feet thick, which had undulated up into our domes. We eagerly submitted the find to the French Academy of Sciences, which published our paper.

Throughout the excitement, one disconcerting fact kept rubbing uncomfortably at the back of my mind. Every oil tycoon knows that salt domes often entrap petroleum. Our domes were located in deep water, some eighty-five hundred feet down. Drilling there would be extraordinarily hazardous and accidents, inevitable. Oil spills from drilling in relatively shallow waters off Santa Barbara and in the Gulf of Mexico have been damaging enough; a spill in the deep Mediterranean would induce nothing short of disaster. The nearly enclosed waters would be renewed only very slowly; oil would irreparably destroy the sea's fragile plant life and threaten its scant animal populations, already reduced by pollution and the effects of tourism. What had I done by publicizing the find? I felt a stabbing pang of remorse as I realized that the information in our article could tempt a profit-hungry bureaucrat to disregard all perils and drill at any environmental cost. Yet I also realized that as a participant in a scientific study—as a citizen of Earth—I was obligated to pursue new knowledge about the planet, to take joy in discovering any one of its innumerable complexities, and not to decree whether society has a right to know about them.

If such choices have been difficult for me, in my brief encounters with science, similar questions have been anguishing for everyone in the human community, relentlessly faced with scientific developments that may result in menacing applications. There was a time when humankind was intoxicated with the raptures of research; today we suffer hangovers as we learn how the discoveries are being abused. We watch biology, the study of life, lead to biological warfare, mechanism of death; chemistry lead to medicines that cure disease and to pollution that induces it; physics, to new concepts of the universe that have liberated the human mind as well as to atomic weapons that have shackled us with the knowledge of an ever-present threat. The scientific means for exalting the human species have been overshadowed by the technological means for eliminating it. Many of us fight the temptation to rail against all those who have called their science "pure," cursing them for having revealed their seductive but deadly little truths in the first place.

Yet cursing scientists for identifying the power of nature is neither very productive nor, in this precarious human epoch, even very sane. No longer is just the quality of life at stake; today the possibility of life itself is threatened. If we want to survive, we must realize that blaming the scientist is no more corrective than condemning the messenger. The paramount decisions of our time concern not what discoveries scientists reveal but the ways those discoveries are used. Yet these are the very decisions that we, the public at large, are forfeiting—surrendering urgent choices about our own fate to a small elite among military and industrial leaders. In most cases, we have not elected the people who are determining our scientific-technological future; we have not selected them on the basis of some special moral training, philosophic vision, or understanding of the past that might guide them in determining our future. But we yield to them, allowing them to decide our destiny on the basis of an overabundance of knowledge and a poverty of wisdom. We have allowed them to change our relation to science from an attempt to learn about nature's might into an attempt to wield it. We have allowed them to narrow the goal of science from a quest for global progress to a quest for national power and personal profit. We have allowed them to make their decisions in secret, applying science *in* our society with no regard *for* our society, no answerability *to* our society. We have allowed ourselves to be intimidated by the experts, finding it easier to plead ignorance of new scientific proposals than to learn about them or to voice our opinions, our needs, our wants, our demands. We have relinquished our human obligation to the future and our civil right to play a role in determining it. We who pay for scientific research, who pay for scientific application, who pay for the weapons and pesticides and poisons—we have allowed science to be taken from us, even used against us. Neither the scientists nor the arms manufacturers nor the profiteers are solely culpable. We ourselves have allowed science in the human community to be sundered from the ethics of the human community. And we alone can and must put it back.

Far from attempting to control science, few among the general public even seem to recognize just what "science" entails. Because lethal technologies seem to spring spontaneously from scientific discoveries, most people regard dangerous technology as no more than the bitter fruit of science, the real root of all evil. Yet today, when so much depends on our informed action, we as voters and taxpayers can no longer afford to confuse science and technology, to confound "pure" science and "applied" science.

The pure scientist discovers the universe. The applied scientist exploits existing scientific discoveries to create a usable product. When I needed an apparatus to help me linger below the surface of the sea, Émile Gagnan and I used well-known scientific principles about compressed gases to invent the Aqualung; we *applied* science. The Aqualung is only a tool. The point of the Aqualung—of the computer, the CAT scan, the vaccine, radar, the rocket, the bomb, and all other applied science—is utility. In total contrast, no motto better suits pure science than Michael Faraday's legendary retort to a prime minister who asked if his newly discovered electromagnetic waves were of any use. "Of what use," Faraday replied, "is a child?" Only the applied scientist sets out to find a "useful" pot of gold. The pure scientist sets out to find nothing. Anything. Everything. The applied scientist is a prospector. The pure scientist is an explorer.

Just as the drive to explore pushes me toward an ever-more-distant horizon, it compels the pure scientist toward the ever-unattainable answer. While the applied scientist longs to achieve a known objective, the pure scientist and the explorer long only to know the unknown. Countless times I've told journalists that our team had no goal, no idea what we would find on a mission; had I as an explorer known what I would find, I would not have gone. So too does the scientist venture into a cell, into the atom, out through the cosmos, unaware of what awaits. The explorer and the pure scientist live only for the search, live only for the hope, live only for the moment expressed even in the Latin roots of the word *exploration*: "to cry aloud on finding." Nobel laureate Peter Medawar spoke for explorers when he described his "compulsion," his "acute discomfort at incomprehension." Nobel laureate I. I. Rabi spoke for explorers when he confessed the *libido sciendi*, the "lust for knowing." Our century's most distinguished philosophers of science describe the pure passion for exploration when they write of their "irresistable need," their "primary urge," their "glorious entertainment," and even of *"laissez faire, laissez jouer."* Let us be; let us play.

Was it science, exploration, or play when we on *Calypso* followed a pregnant, migrating whale, just to see if she slept at night? When we lowered a rag-stuffed mannequin in shark-infested waters—to see if the creatures would bite a lifeless, bloodless, scentless doll—were we motivated by science, exploration, or play? Our adventures involved each of the three and all of the three: Exploration and play beget pure science. Of what practical "use" is it to know if a pregnant whale sleeps? Probably none. It does not matter. What matters is that the human species is gifted not just with eyes to see nature but also with a mind to glean its significance. Science, exploration, and play: To learn is the nature of the child; to seek is the necessity of the explorer. But to learn, to seek, to understand—this is the mission of the pure scientist.

Sleepwalkers is the word Arthur Koestler used to characterize scientists as they meandered from theory to sidetrack until finally they chanced to arrive before some grand new understanding. Yet their traditionally oblivious nature is precisely the characteristic that angers those who indict researchers as being responsible for our present technological predicaments. Our beautiful dreamers too often wander in the universe and realize—only after they have gone too far to turn back—that their ramblings have led to destructive nightmares they never intended or even imagined. The harshest critics ask scientists questions that seem justifiable: At what point does careless inattentiveness become negligence? Is not negligence itself a crime, and negligence that endangers human lives and even human life itself the ultimate crime? Should we not demand that scientists become alert to impending dangers? Shouldn't we awaken the sleepwalkers?

However logical these questions seem, the backlight of history reveals the absurdity of expecting scientists to be soothsayers, of demanding that they predict the outcome of their work. Which of the scientists who appear so guilty today could have acted on specific premonitions to alter their actions in decades past? The tragic course of radioactivity research proves the point. Could our fearsome atomic arena have been prevented by Robert Millikan, founder of Caltech, who twenty years before the bomb declared confidently that "the idea of there being 'dangerous' quantities of subatomic energy, like most hobgoblins that crowd in on the mind of ignorance," is "a myth"? Could the bomb have been averted by Ernest Rutherford, the pioneering physicist often compared to Newton, who assured his followers that "anyone who expects a source of power from the transformation of these atoms is talking moonshine"? Could our predicament have been precluded by some special vigilance on the part of Albert Einstein, whose theory of relativity revealed the powerful relation between matter and energy but who, just six years before the bomb was dropped, was asked about the possibility of uranium fission and replied in surprise, "That never occurred to me"? Should we today blame Otto Hahn, who chanced upon the discovery of nuclear fission but remained so unsuspecting of its implications that, when Hiroshima was incinerated in the firestorms ignited by his theory, his friends feared that the grieving physicist would take his own life? He certainly blamed himself.

"Here's to pure mathematics, may it never find an application," the renowned mathematician G. H. Hardy once toasted. Yet the remark that seemed wistful then seems only ironic today: Mathematics now gives technologists the ability to aim missiles, with computer precision, directly at cities.

"Comprehension is the enemy of action," Nietzsche once wrote. I have felt the sting of truth in that remark as I have watched it apply precisely even to my own small world. Today I see what profiteering SCUBA divers do with the Aqualung—using it to enable themselves to shatter coral reefs and sell the fragments as souvenirs; to scour underwater grottos of every last fish, snatching all creatures out of the hiding places in which they had escaped fishermen's nets. Now that I understand, I am not sure that the good the apparatus can do outweighs the bad. Could I turn back time, I do not know whether I would participate in the invention of the Aqualung again.

Many social commentators and even some pure scientists themselves, jolted too often by the bruising recoil of apparently innocuous discoveries, have also reacted according to Nietzsche's dictum. Several microbiologists have abandoned their careers with a public declaration that they did not trust those in power who one day might apply their findings. In the *New York Times*, futurologist Herman Kahn once called for an "index of forbidden knowledge," his own throwback to the Vatican's sixteenth-century *Index of Forbidden Books*. Others have proposed preventing the advance of pure science by cutting off grants for research and even by confiscating essential equipment, including the electron microscope. To these despairing souls, the outright censorship of science seems like setting a counterfire—igniting a second blaze to combat the first.

Yet those who so dread the consequences of science that they would willingly abridge the essential freedom to learn ignore one ultimately unavoidable fact: While censorship itself is oppressive and totalitarian, censorship of the universe is simply impossible.

As books have burned, only pages have turned to ashes, not ideas. So certainly it goes with the universe. What one person does not discover, another one will: History proves this point with its long list of instances in which one discovery was revealed simultaneously by several scientists, each working independently in distant countries. No one can succeed in censoring knowledge because no one can negate existence. Closing our eyes to nature simply makes us blind men as we enter a fool's paradise. When illustrious seventeenth-century scientists refused to look through Galileo's telescope, they did not stop the moons of Jupiter from spinning—or society from eventually finding out about them. When an early physician, also described by Galileo, once gazed at a human body that a colleague had autopsied, admitted that he saw the cables of nerves that emanated from brain to limbs, and still insisted on clinging to the ancient belief that the heart rules thought and action, not one single brain ceased to initiate movement; not one single heart began to think.

"We can easily forgive a child who fears the dark," Plato wrote. "The real tragedy of life is when men fear the light." The desperate notion that it is somehow easier to keep the whole cosmos a secret than to control the few human beings who abuse it reveals how helpless and hopeless the public feels. We don't need to censor science; we need to censure the men who manipulate it to harmful ends. The history of how science came to be abused is no more mysterious than the history of how the human being came to be abusive.

Science was born in a spiritual arena. Awed by the wonders of the earth, humans first explained nature in the way that seemed most obviously apt—as enchanted and ruled by gods. Even as the millennia sped by, through the Middle Ages and up to the Renaissance, science remained the province of humans impassioned by the spectacle of a universe they cherished as a supreme gift. Monasteries were islands of scientific appreciation, and scientists drew on all the glories of human existence—music, humanities, the arts, faith—to arrive at, and enrich, their understanding.

This is not to suggest, of course, that human nature itself was different in those early eras. From the beginning, humans abused the powers they perceived as so miraculous. One of the first natural phenomena to be explained—the secret of the eclipse—was used by Babylonian priests to terrify their followers. Archimedes and Leonardo both solicited funds for their research by advertising their ingenuity in war making. Galileo most likely hit on his idea for calculating the trajectories of planets when he was paid to calculate the trajectories of cannonballs. The difference between the misdeeds of science past and those of science present lies only in their degrees of success.

Yet both scientists' fundamental reverence for life, as well as the relative destructiveness of their frequent lapses, were to change through the eras. The first landmark in science—when researchers began to lose their sense of divine inspiration—probably did not abruptly appear, as legend holds, because of a sudden divorce between church and science at Galileo's trial for heresy. Most of the scientists of the time were deeply religious men; Kepler, for example, who discovered the speeds of planets by relating their orbits to harmonies, poetically called himself "God's witness." Galileo himself maintained that, while monks interpreted the written Bible, he was attempting only to read God's living Bible.

The alterations that made science seem more fearsome may have occurred, ironically, because by the time that people recognized scientists as eyewitnesses of creation, the scientists themselves were backing away from their role as spokesmen for its wonders. Hoping to arrive at truths more efficiently, Galileo and his colleagues began to follow the "experimental method," banishing everything extraneous from their observations—sound, color, odor, taste, emotion. With their telescopes, microscopes, and firsthand visions of heaven, scientists were able to replace blind faith with bare fact. They ordained themselves into a new kind of priesthood. Yet the heavens that these new visionaries saw were only empty infinities. With the seventeenth-century scientific revolution—the advent of Galileo's mechanical experimental method and the concerted effort to research clinically rather than to let emotion influence findings— scientists were expected to approach nature in the way a mortician approaches the corpse he embalms. Many of them perceived a great deal about the body of science but little about its soul.

Bestowed with new authority, scientists grew to admire the human intellect more than the phenomena that intellect enabled them to discover. About the time that Galileo conceived of his emotionless experimental method, Bacon was declaring "knowledge is power." Within only a few years, Descartes boasted that this new "practical" method of science would "render ourselves lords and possessors of nature." By the time a century had passed, Kant was proclaiming, "Give me matter and I will build a world from it." Hermann Helmholtz, one of the greatest physiologists of the time, scoffed that "if a human eye had been sent to me to test as an instrument for the reflection and refraction of light, I would have sent it back to the maker."

These approaches set a potentially dangerous new scientific stage. Yet, aside from occasional forays into weapons making, the fundamentally good intentions of scientists remained the same for centuries. Most scientists and philosophers—Bacon, Descartes, and Kant foremost among them—prophesied that scientists would use their limitless new capacities to build a better world for all humanity. For the time being, however, only their dreams were big. Their abilities stayed small. Even as late as the nineteenth century, it simply took too long—up to fifty or a hundred years—for scientists to turn a discovery into a technological reality, be it for good or evil.

The gap between discovery and technology closed—scientists learned to apply their findings immediately to inventions—just at the outbreak of World War II. Humanity acquired its cosmic powers exactly at the moment when nations chafed to wield them against one another. The world armies posted science's second landmark: They appropriated the scientific quest for power and altered it into a military quest for supremacy. The world erupted into the battle Winston Churchill dubbed the "Wizard War."

Our wizards mobilized themselves, personally offering their services to the military. Even with hindsight, the world understands—I understand—the scientists' originally patriotic motives. They responded to the exigencies of war. Yet the relevant question for us now does not concern the reasons why scientists offered their powers to the military; more important are the permanent changes the military in turn made in the way we use science.

With the authority traditionally granted to the armed forces in wartime, the military first wrested control of all scientific findings from the scientists who had discovered them. After Harold Urey identified "heavy water," thought at the time to be essential to fission, he was blocked from all information about the methods his employer, the DuPont company, was using to exploit his discovery. Scientists were so completely severed from decision making that the use of the atom bomb was approved by leaders who had never even been told that radiation causes genetic damage. Neither Churchill, nor the British prime minister who concurred with Truman's decision to target a population, nor, as far as history shows, *Truman himself* knew that the nuclear weapon would levy effects beyond the deaths caused by its initial explosion. Herman Muller had published his findings about radiation's genetic effects eighteen years before; but biologists, who would have raised objections, had no idea that a bomb was even being built. Elected leaders approved the use of atomic weapons by default—on partial information. Only the military had all the pieces; in effect, only the military made the decisions. The scientist as priest had been deposed.

Once military officials had installed themselves as the new patriarchs of science, how did they change our attitude toward science, the search of the universe, as well as toward scientists, explorers of the unknown? "To war-time decision makers," I. I. Rabi—the scientist whose theories led to the microwave, the maser, and the laser—once bitterly remarked, "the scientist is like a trained monkey who goes up the coconut tree to bring down choice coconuts." The military, in short, demoted scientists to draftees. The scientists who were deemed acceptable for military work were neither masters of their discoveries nor masters even of their own conscience. Those endorsed and promoted in the war-time hierarchy espoused the values described by physicist Robert Oppenheimer, father of the atomic bomb. "When you see something that is technically sweet," Oppenheimer once remarked, "you go ahead and do it. And you argue about what to do about it only after you have had your technical success."

I understand the "exigencies" of war. I fought in World War II; I fought for my country and I fought out of patriotism. Nonetheless an anguishing vision burns in my memory. I sailed on the cruiser *Dupleix* with two other cruisers, escorted by torpedo boats, to the coast of Italy. One night, off Genoa, I listened to the sounds of guns, the sounds of destruction, the sounds of death. Not only soldiers or sailors were dying in Genoa that night. Townspeople were dying. Wives and husbands and children were dying, as nations acted on what had seemed the unthinkable notion of killing concentrated populations of noncombatants. My ship was part of the attacking force. I listened to the sounds but I did not watch the sights. The officer directing our fire from atop the gunning tower announced, over a loudspeaker, the point of impact of each of our shots. At our third salvo, he announced: on target. We understood. And together we wept.

Radar, the proximity fuse, sonar, the automatic control of gunfire—these, the unholy offspring of the coupling of science and technology, allowed the world's militaries to kill families with devastating accuracy in the dark, as they slept. Were they created because of the exigencies of war alone? Or were they created partly because some scientist had a technological sweet tooth? When Oppenheimer finally made his now-renowned confession that physicists during the war had "known sin," scientists involved in his atom bomb project indignantly disassociated themselves from his comment, repudiating any sense of remorse. Free-man Dyson, the great theoretical physicist known for his work in quantum mechanics and theories about space exploration, became acquainted with many of these men after the war and noted the adolescent relish with which they reminisced about their bomb-building days. "The sin of the physicists at Los Alamos did not lie in their having built a lethal weapon. To have built the bomb, when their country was engaged in a desperate war against Hitler's Germany, was morally justifiable," Dyson wrote. "But they enjoyed building it. They had the time of their lives while they were building it. That, I believe, is what Oppy had in mind when he said they had sinned. And he was right."

To the military martinet, this kind of attitude made a scientist the perfect marionette. Researchers were no longer regarded as people but as possessions—no more, no less, than superweapons themselves. When the war in Europe was over, did the French, Americans, British, and So-viets swarm over German borders to claim national treasures, or jewels, or land, or the other traditional spoils of war? In fact, they scrambled to collect German scientists, actually labeling these human beings as "war booty" and "intellectual reparations." The American military even awarded German scientists with U.S. citizenship in exchange for help in plotting against the Japanese; only two months before, German scientists had guided the Japanese through their missile factories, helping them plot against the Americans. Principle was sacrificed to purpose. Antithetical and even convertible patriotism was overlooked in the obsessive quest for the biggest bang, even when the war was over.

Industry inflicted the final malformation on our science in those postwar years. To the military's quest for power, business has added its own driving quest for profit. The combination of science and industry, of course, is not necessarily an evil. Even at the start of the industrial revolution, some of the world's great scientists used their genius to bring about an economic and altruistic gain for all society. In the 1860s, Pasteur not only applied his germ theory to create "Pasteurization," rescuing France's wine and vinegar industries, but also found both the cause and cure of silkworm disease, saving growers millions of dollars. When Napoleon asked the scientist why he had not legitimately profited by his findings, Pasteur replied: "In France scientists would consider they lowered themselves by doing so."

Yet the fact that science was not exploited for profit in that era was probably again due less to such generosity of human spirit than to the lengthy time lag that separated discoveries and ideas for their application. The period from the industrial revolution up to World War II was one in which industry recognized how profitable the application of knowledge could be. A scientific discovery could, within months, produce not only the atom bomb but also penicillin, plasma, and the first electronic computer. Science, chief executives realized, could be applied to increase production and decrease cost, to create new goods that were not necessarily durable but certainly salable. Industry's stock soared in exact proportion to its expenditure on science. Even a country's gross national product reflected its gross national success in applied science.

No one would deny that the marriage between science and industry strengthened the world's economy. But what kind of an economy has this kind of exploitation of our science left us? Weapons production accounts for one of the largest single portions of most developed nations' gross national products. Arms manufacture increases employment and pours wages into an economy without introducing consumer goods to compete with other products. Even better for dividends, weapon stockpiles continually become obsolete and must be completely replaced. Social services, of course, could offer the same benefits—increasing employment without introducing competing consumer goods. Instead, industry and military use our science to lock us into a *Wehrwirtshaft*—a "preparedness economy"—one in which our entire human community is engaged in, with our needs subordinate to, the production of arms. Gerard Piel, owner of the *Scientific American* and president of the American Association for the Advancement of Science, both chronicled and protested these developments. Piel noted that we have used science, the sum of human knowledge, to create the kind of society in which a New York newspaper headline once actually blared: "Fear of Peace Depresses Market!"

Such squandering of our intellectual resources from science has been virtually institutionalized in recent years, due to two specific public policies: misallocation of research funds and scientific secrecy.

First, governments use our taxes to support applied science almost exclusively, nearly choking off all progress in pure, undirected exploration of the universe. Because I remember the lavish funding of pure research immediately after the war, I all too keenly understand the damage caused when those funds were cut. The length of women's skirts is decided in Paris; the tone of world science is dictated by Washington, D.C. The United States opened a golden era of scientific discovery in the forties. The U.S. Office of Naval Research alone is probably responsible for enlarging the scope of human knowledge more than any other single institution during my time. I remember many interesting studies of seafloor topography funded by the agency. It also provided ample funds, with no military strings attached, to scientists curious about natural mysteries that had no connection whatsoever with the oceans. Scientists from all over the world flocked to the United States. During those years, I met an intelligent young man in France who dreamed of becoming an oceanographer. I helped him obtain a scholarship at Scripps Oceanographic Institute in San Diego. Six months later, I traveled to California on business, invited him to dinner, and asked how he was doing. The young man paused before he answered. When he finally responded, he looked at me balefully. "Here, I am confronted with my own limitations."

I was struck by the philosophic way in which he realized that he was bounded by the finite extent of his own abilities. In an atmosphere of lavish funding that bought everything a scientist could need, a student had only himself to blame if he failed to discover something extraordinary.

The U.S. Pentagon blocked this otherwise unstoppable advance toward the future in 1963, with a document characterized by such lack of forward vision that its title—"Operation Hindsight"—turned out to be unwittingly apt. The study contended that the massive funding the American military had devoted to pure, exploratory science had not paid off in proportionate advances in weaponry. The U.S. Congress gave this nearsightedness the force of law, passing the Mansfield Amendment, which forbid the military to fund any scientific inquiry without direct military consequences.

Because most of the remaining funds for science came from private industry, which gives funds almost exclusively to applied research, pure science in America was left with very little nourishment. In the intervening years, the situation has only deteriorated. Today more than 72 percent of the entire U.S. government science budget is devoted to weapons research. About 25 percent of the total federal U.S. science budget goes to applied science. Pure research, unfettered exploration, receives only 3 or 4 percent of the funds that the American government spends on science.

The migration of scientists I once saw heading to America has turned into an emigration out of America. Yet scientists leave to no avail. The nations of the world have followed the deadening scientific strategies of the United States. Almost half of all the money we in the world spend on science today supports the development of more weapons. There is an irony in this unbalanced budget: Devoting such exorbitant sums to weapons research will not necessarily enhance national security. And while spending nearly all the rest of our research funds on applied science might sharpen a nation's industrial competitive edge in the short term, in the end this kind of budget will ultimately, and unavoidably, impede progress.

The simple truth: Applied research generates improvements, not breakthroughs. Great scientific advances spring from pure research. Even scientists renowned for their "useful" applied discoveries often achieved success only when they abandoned their ostensible applied-science goal and allowed their minds to soar—as when Alexander Fleming, "just playing about," refrained from throwing away the green molds that had ruined his experiment, studied them, and discovered penicillin. Or when C. A. Clarke, a physician affiliated with the University of Liverpool, became intrigued in the 1950s by the genetically created color patterns that emerged when he cross-bred butterflies as a hobby. His fascination led him—"by the pleasant route of pursuing idle curiosity"—to the successful idea for preventing the sometimes fatal anemia that threatened babies born of a positive-Rhesus-factor father and a negative-Rhesus-factor mother.

With their strict applied-research rules, governments today command, "Don't search. Find." But what in the world are scientists supposed to find when they have no idea what they are looking for? By initiating the Age of Goalism, we are ending the Age of the Dreamer.

More and more, scientists who simply wish to pursue their curiosity, as Clarke did, must camouflage their goals. The scientist who wants to count the legs of a bug can get funding only by saying he plans a pesticide project. The politics of gaining a grant force the scientist to subordinate his goal of gaining knowledge. Not long ago oceanographers around the world eagerly awaited a satellite proposed by American scientists. Seasat, as it was called, would orbit the globe with a capacity for collecting previously undreamed-of information about the world's waters. I accepted several invitations to visit the Jet Propulsion Laboratory to assess the progress of the program. The scientists had proposed some very interesting instruments, but they had limited the satellite's purview to just the seas adjacent to America. Why cripple the project this way, when humans know so little about the open ocean? The scientists told me that the U.S. Congress, acting on what it felt to be public desires, would fund only programs aimed at U.S. waters. The sense of that argument is nonsense: This satellite would spin around the world, traveling over open oceans, passing above coasts of all nations, and could have enlightened the human community with a fantastic amount of new information. But because politicians do not want other nations to benefit from new knowledge, they deprive themselves of the knowledge as well. At any rate, Seasat was apparently cursed by more than its creators' shortsightedness: Soon after it was launched, it stopped working.

The fundamental mistake here lies in the assertion that the *public* wants only "useful" science; that the *public* disdains all research that would garner knowledge or understanding. I don't remember any time the public declared that knowledge in itself is impractical. When did the public announce that "progress" entails only industrial progress, only military progress, and not intellectual progress? When did the public deliver the edict that we must broaden only our economies, never our minds? Industry, the military, and through them politicians certainly share greatly in the plaudits that come if profits accrue, in the short term, as a result of funding applied science. Yet it is all of us who bear the full brunt of the long-term intellectual losses incurred from stinting on pure science. And those losses are incalculable.

In 1970, for example, I accepted a request from the Bechtel Corporation to take four ships, including *Calypso*, to survey a projected path for a natural-gas pipeline to lead from Tunisia up the coast of Africa and across the Mediterranean to Sicily. We began by using sonar to map the general topography of the bottom; and then, day after day, my friend and *Calypso* colleague Albert Falco took our diving saucer down into the deep to check firsthand for problematic areas, dictating the details of the seafloor into a tape recorder.

On October 30 of that year, Falco was scouting deep waters, propelling the diving saucer farther and farther down along the sloping bottom. When he had descended about 750 feet below the surface, several dark shapes jutted up out of the sands and into the glow of the saucer's searchlights. Ancient amphorae! He swerved the saucer around and passed over the litter of scattered wine and water jugs again. There was absolutely no doubt in his mind, he said into the recorder: The remains of a ship lay engulfed in the muds. Through the years the wooden hull planks had rotted away; fragments of ancient pottery, together with pieces of the ship's metal stores and gears that had resisted corrosion, had settled neatly to the bottom, outlining the original shape of the vanishing ship.

Yet there was something unusual about this venerable ghost, profiled by its heap of cargo. Its shape was much longer and far more narrow than all the ancient Roman ships that Falco had ever seen. He maneuvered the saucer closer, hovering just above the wreckage. The ship's fittings were forged of bronze. Sleek . . . capable of great speeds . . . richly ornamented—this was no merchant ship. Falco realized he was gazing upon a war vessel, an extremely ancient man-of-war, at that. The Phoenicians—the Bible's "Canaanites"—are known to have built such long, narrow warships; but only a few drawings and embossed Assyrian coins, most of which by now have disappeared, have shown our modern world how the ships appeared, without providing much detail. Suddenly Falco recognized a long, menacing form that loomed in the shadows—a battering ram! Of solid bronze. The Phoenicians, too, were renowned as the first people to fit the huge, deadly weapon into their ships.

This sunken vessel was no mere monument to modern tragedy, like the ironclad *Monitor* off the coast of Virginia or the *Titanic* in the Atlantic. The ship, Falco knew, was a time capsule from a civilization whose culture long ago began to fade from the record of human knowledge. He spoke wistfully into the saucer's tape recorder: How he longed to stay at the spot, to abandon the pipeline project altogether. Yet his obligations reduced him to using the saucer's mechanical claw only to seize a small, unidentifiable piece of bronze, which he would bring back to the surface later, after his day's work was finished.

If in fact the vessel Falco found was Phoenician, it came from a civilization that was sailing ships two thousand years before a few wandering tribesmen even settled on the hilltop that one day would be Rome—from a civilization that ruled the entire eastern Mediterranean by the time King Solomon reigned over the biblical empire of Israel. Had we been working on a project in another kind of world—admittedly utopian but certainly more reasonable than ours—such an extraordinary discovery would have brought work on the pipeline to an abrupt halt, and we could have devoted all the expedition's energies and resources to an exhaustive study of this cultural treasure. Yet all we took from the spot was the piece of bronze, a photo, and the coordinates, which I still have in my log. We had to leave our dreams behind. In the real world, only the funds to survey a pipeline path, for more gas to be burned away in Sicily, were readily available. Who would pay for a discovery that would not momentarily light a stove, but could instead forever illuminate a lost civilization?

Even more harmful than such funding travesties are policies of systematic secrecy around pure science. Secrecy deprives the public not only of knowledge but in some cases even of day-to-day safety. Citing the pretext of national security, defense departments and intelligence agencies have long shoved thousands of scientific papers concerning applied research into the classified files; but recently military officials have demanded that unclassified pure research conducted with their funds be kept strictly within national bounds. Government security officers have ordered some researchers to present their findings at science conferences only when foreigners are barred from the room. Driven by competition, scientists now even censor themselves; international meetings of scientists engaged in industrial research no longer resound, as I remember they once did, with excited and spontaneous conversations. The fear of losing some possible future patent now settles over the room like a cloud. No one seems to bargain for the inevitable outcome: As one man withholds his discoveries, others will withhold discoveries from him. His own progress is obstructed, along with scientific progress in general.

While those who monopolize their findings for profit may seem miserly, anyone who withholds findings that could specifically affect human safety, health, or life is nothing less than a criminal. In many civilized countries, anyone who refrains from helping an endangered person can be subject to arrest and prosecution. Are executives who, in the name of competition, withhold information about a new life-saving drug so different from the bystander who watches people die while he has the power to help them stay alive? Does anyone who discovers a method to save lives "own" that method?

I have watched companies claim precisely such proprietary rights as they patent information vital to the safety of divers. Even after Gagnan and I had invented the Aqualung, I dreamed of finding yet another way of diving, one that would allow me to go still deeper, stay below still longer. I had once even envisioned colonies of undersea villages on the continental shelf. In 1957, Captain George Bond, an American Navy doctor, conceived of "saturation diving," just the physiological theory I needed. For a time, the U.S. Navy had expressed little interest; he came to me, and I eagerly joined in his project. Bond proposed that divers eventually become "saturated" with nitrogen and can continue to stay in the depths without further increasing their decompression time, coming and going from habitats filled with pressurized air. When their work is finished, they can be hauled up, comfortable in their habitats, for one marathon decompression session.

In 1962, during the first of what we called our "Conshelf" experiments with saturation diving, Falco and another member of my crew, Claude Wesly, spent a week living 33 feet below in a specially compressed chamber, leaving for "work" each day at 80 feet. The next year, we went a step further. We built an undersea village, complete with a compressed-air dormitory and even a diving-saucer garage, which we submerged among the sharks that cut through the waters of the Shab Rumi coral reef in the Red Sea. Our oceanauts lived there a full month, filming life among the creatures of the reef for my film *World Without Sun*. Finally, in 1965, we immersed a 130-ton underwater house in deep waters, 330 feet below the surface of the Mediterranean off Cap Ferrat on the Riviera. Six oceanauts, including my son Philippe, lived at this great depth for three weeks, leaving their undersea house each day to toil on a mock oil rig at 400 feet. Eventually they perfected their abilities to the point that they could perform tasks below as quickly as rig workers finish them in oil fields.

Thanks to their daring, divers today can more safely control oil exploitation in deep waters; salvage divers can retrieve barrels of toxins scattered in deep waters by wrecks of massive cargo ships; aquafarmers can better tend and harvest their crops. We released all our findings about deep-diving physiology. Our expeditions, however, were just pioneering in the field. Many refinements were left to be made. Corporate industry has since, at great cost, worked on saturation diving, improving gas mixtures for breathing and reducing decompression time, and thereby increasing the efficiency of their divers' underwater work. Yet, driven by competition with other corporations, they have then clamped tight the lid of secrecy on all this information so vital to divers' safety. They treat divers' lives as negotiable and negligible commodities. They claim monopoly rights to breath itself.

But the secrecy trend does not stop at censorship of advances that protect life; politicians, industry, and the military censor dangers that threaten life as well. They do not "fund" science; they "buy" science. Governments review the safety of defoliants, food additives, leaded gasoline; they hire scientists paid by the industries under review and thereby purchase exactly the results most expedient for their plans. When tests could not be tilted so as to avoid a damaging truth, government agencies have sometimes contended that, since they have "bought" the science, they "own" the results and can refuse to release them— never mind that they have "bought" those results with our own tax money. By buying facts, agencies buy national policy; they force elected representatives to base law on half-truths or no truths at all. Not even a president is immune to this emasculation. Once I went to the Élysée Palace to discuss some specifics about national ocean policy with President Georges Pompidou. When I had spoken for my allotted fifteen minutes, he gave no indication that my time was up; after I had continued a while more, the president began to speak, agreeing with the statements I'd made and even elaborating on arguments that further supported the important points. Finally, Pompidou shook my hand and bade me farewell. Then he concluded with one sentence I can still repeat, word for word, today. "But never forget, Cousteau: The president of the French republic can do nothing against his powerful agencies." (I suspected, of course, that he meant Électricité de France, Energie Atomique, and CNEXO, the national center for ocean exploration.)

Françoise Giroud, a former French minister of women's equality, neatly summarized the situation in her memoir, *The Comedy of Power*. "Government," she wrote, "is the art of keeping people from meddling in their own business."

What happens when the people who preside over science have unlimited funds and an unlimited capacity for keeping their actions secret? They reduce science policy to a crude policy of anything goes. They brandish the power of nature, whatever the consequences for Earth in general or for humanity in particular. Their heedlessness reminds me of the way my son Philippe behaved when he was five. He found a rubber band, a stick, and a stone; he saw a bird. He had all the elements necessary for taking a shot, and so he did. The bird thumped to the ground, dead. After an instant of stunned silence, Philippe ran to the house shrieking with remorse, knowing even then that his tears came too late. Those who wield science against our society, of course, are not five-year-olds dealing with sticks and stones and rubber bands. Nonetheless, as soon as a technology becomes possible, they take their shot.

○ When airlines proposed building supersonic transports (SSTs) for commercial travel, experts barraged governments with protests, fearing that a fleet of such planes could destroy the planet's protective ozone layer, exposing Earth to greater levels of ultraviolet radiation and increasing the rate of skin cancer. Though French technologists scorned all dangers and proceeded, the fact that Americans abandoned building their proposed sixteen planes led the public to suppose it had won a victory. Nonetheless, hundreds of supersonic military fighter planes, including those of the U.S. Air Force, continued to incinerate our ozone unopposed.

○ Immediately on discovery of the Van Allen belt, a zone of charged particles around the earth, with mysterious but intricate effects on all life, technologists decided to explode a hydrogen bomb in the phenomenon to see if they could produce a colorful aurora by disturbing the particles. They shrugged away horrified opposition from pure scientists who objected to such foolhardy meddling with the unknown and—neither knowing nor caring what harm they might inflict on living beings—they set off the explosive that they advertised as the "Rainbow Bomb."

Contemplating such heedless arrogance on the part of applied scientists and technologists, more than one pure scientist has responded with the Duke of Wellington's description of his own generals: "They may not frighten the enemy, but by God, they frighten me." A relatively minuscule coterie of technologists rule over the scientific innovations that affect our lives. They think they control science. In fact, science is out of control.

It is ironic and perhaps significant that, as our social priorities have changed through the centuries, the legend of Faust has changed. Today we tell a tale about a Faust who so hungered for power that he sold his soul to the devil to get it. Yet in the original account of Doctor Faustus—in 1587, back in the days when Descartes and Bacon dreamed of using science to help humankind—the magician did not ask for power at all: He asked for plentiful food and clothing, for pocket change, and for a way in which he could fly through the stars. We have evolved in the same way as poor Faust. We once hoped to use science to achieve exactly the same lofty aims—to help *all* humanity. Instead we have so completely sold our souls and our science for power that when a physics professor at a leading university once asked his colleagues to take a mild Hippocratic oath of physics in front of their students—"I will use my knowledge of physics for the benefit of society rather than for harm"—he was not only forced to retract the statement but even nearly fired.

Yet are our science decision makers—those who preside over science budgets and those who bless or condemn proposals—solely responsible for the sins committed in the name of science, or do we, in our own complacency, share the guilt? Every day new scientific issues arise that cry out for ethical judgments and that very much involve our own personal well-being. Dilemmas abound: Some ecologists have proposed a "triage" list for endangered species, which would assign protection priority to animals based partly on their supposed "importance" to humans. Should we condemn such arbitrary evaluations of life as meaningless and even immoral; or should we let endangered species die at random? By rearranging bits of genetic material, we can reprogram heritage, turning a bacterium, for example, from an organism that produces toxins into one that produces medicines, like interferon. Yet this technique could conceivably be abused to create new and virulent diseases as well. True progress is achieved not by mindless leaps into the unknown, but by enlightened choices. What should we do?

Who shall respond? The stockholders and inventors who stand most to profit from a breakneck plunge into development? The militaries and industries that will turn all discoveries into weapons inventions? We know by now humanity will not progress with another, better bomb; we need a better idea, a higher ideal. Should we turn, then, to the churches? The faithful were in fact given a clear moral directive for science when Pope John Paul II, addressing the Pontifical Academy of Science on November 12, 1983, courageously instructed scientists to turn their backs on weapons building: "By refusing certain fields of research inevitably destined, in the concrete historical circumstances, for deadly purposes, the scientists of the whole world ought to be united in a common readiness to disarm science and to form a providential force for peace." Yet only four days before, the bishops of France had exempted their own countrymen from exactly this ethical imperative, ruling that nuclear deterrence is a legitimate defense against "Marxist-Leninist ideology."

The accelerating advance of science and technology, unfortunately, will not stop as we try to straighten out such dogmatic contradictions. We have paused before, as decisions awaited, and the decisions were made without us. The questions posed by new scientific capabilities are indeed anguishing ones. We, the public, may not have all the answers, but we have common sense.

In his classic work *The Two Cultures*, C. P. Snow provoked the ire of both scientists and humanists when he accused them all of segregating themselves into two opposing factions. But indeed, to fail to work together is a fatal error. If we are to live in safety on our single globe, we must recognize that we need scientists to give us truth and humanitarians to tell us its significance. We need scientists to protect us from nature; we need humanitarians to protect us from ourselves.

Science and poetry are, in fact, inseparable. By providing a vision of life, of Earth, of the universe in all its splendor, science does not challenge human values; it can inspire human values. It does not negate faith; it celebrates faith. When scientists discovered that liquid water, which brought forth life on Earth, exists nowhere else in great quantities in the solar system, the most significant lesson they taught was not that water, or the life that depends on it, is necessarily the result of some chemical accident in space; their most important revelation was that water is rare in infinity, that we should prize it, preserve it, conserve it. When scientists theorized that some microscopic piece of moss may grow on a distant planet, they did not diminish our Earth by wresting away its status as the only place where life may survive; they distinguished it by underscoring once again that our wet, blue globe is probably the only place where *human* life has thrived. When scientists discovered that each living being on Earth—each plant, insect, and animal—supports and in turn depends on all others, they did not reduce creation to a simplistic, mechanistic order; they rather exalted it as a complex, interdependent one. They showed us that when we safeguard our environment, we safeguard ourselves. The Latin words for "with science," after all, serve as the roots for our word *conscience*.

The reason I have made films about the undersea lies simply in my belief that people will protect what they love. Yet we love only what we know. Learning science, learning about nature, is more than the mere right of taxpayers; it is more than the mere responsibility of voters. It is the privilege of the human being.

Science bestows yet another gift, one with even more immediate value than human ennoblement. Science has revealed to us that, in the infinite span of time and space, the similarities shared by all human beings carry far more significance than any differences between individuals. Of all the billions of genetic commands our bodies obey, 99 percent are identical to an ape's. Just 1 percent of our genetic makeup distinguishes us as human. The infinitesimal fraction of genetic material that distinguishes one human from another is negligible. The genes that distinguish a human who lives in one country from a human who lives in another simply do not exist.

The visions science bestows can guide us into a bright new age. But first social science must advance out of the dark ages; first we must stop using the powers of science to threaten other nations and begin using them to heal our global divisions. We must finally learn that one nation's turning against another is like a hand turning against a heart, two living elements of one living body, trying to damage its own parts and thereby destroying its own whole. Humanity's enemy has never been science, or even scientific discoveries concerning arsenic or the atom. Humanity's only enemy is the human being. We do not need to civilize science; we need to use science to civilize civilization.

Humanity has already glimpsed science's glittering potential. We have used science to lengthen the course of the human lifetime; to increase the infant survival rate; to distribute fresh food quickly; to create medicine both to cure and prevent disease. But we have used science to provide these benefits to a fraction of the people on Earth. The scientific-industrial revolution has not been a revolution at all. What kind of revolution fails to provide starving people with sufficient food and instead provides wealthy people with food to waste? We must admit a shocking truth: For more than a half century we have possessed the scientific capability to end world suffering. For half a century we could have used science to eliminate human want. Instead, we have left humans wanting.

The opportunity remains. We can use science for transient financial gain, or we can use science for lasting human gain. We can try to survive despite science, or we can flourish because of it. We can still fulfill Faust's wish. We can feed the starving, clothe the poor, not by making a pact with the devil, but by drawing instead on all that is good and beneficent in the human spirit.

Albert I of Monaco, a great seaman, an inspired oceanographer, and a generous patron of pure science, established the Oceanographic Museum in Monaco in 1910. Having searched his archives, having read his letters, now stored in palace vaults, I almost feel I know him. Albert believed that science could dispel the darkness of superstition, that it could bring peace to Earth. Despite unrelenting demands on his time, he would agree to lecture in Paris, squeezing the trip into his schedule by setting off from Monaco at night on his motorcycle, arriving in the French capital by dawn. He would speak to workers in the early morning, expounding on his dreams of new scientific discoveries that would show human beings that their divisions are artificial, that their unity is real.

Albert's closest friend was Kaiser Wilhelm, king of Prussia. Albert so valued this friendship that, when he opened his prized oceanographic museum to the public, he invited Wilhelm to dedicate the institution.

Four years later, Wilhelm declared war against France, igniting World War I. Not long after the war was over, and Germany had settled only into an uneasy peace, Albert died. I always believed he died of a broken heart.

Yet I believe, as well, that Prince Albert's hopes were not defeated, that they were simply postponed. I believe that Albert was correct; that he was mistaken only in thinking that science could have saved the world within his generation. Today the time is right because time is running out. Desperadoes have been known to triumph in the face of a desperate cause. We have succeeded in using science to conquer nature. Surely we can use science to conquer human nature.

CHAPTER NINE

THE HOT PEACE:
NUCLEAR WEAPONS AND
NUCLEAR ENERGY

I had no idea, as I walked toward the dais at the International Atomic Energy Agency's 1959 conference, that the events to follow would change my life. As director of the Oceanographic Museum in Monaco, I had been in a position to offer our facilities to the new nuclear agency for a meeting in which four hundred biologists, chemists, and physicists, from dozens of nations, planned to participate. In those days, I had little more information about atomic energy than any other average newspaper reader. I drew pleasure from receiving the scientists simply because I recognized their urge to seek, to learn, to discover.

No difference, I thought, between microscopes and Aqualungs, cyclotrons and goggles. They all bestow the freedom to wander where curiosity leads. My exploration tools had allowed me to see extraordinary sights undersea in those early years—who could lay eyes on such mysteries without longing for more? I had searched the night sea for the nautilus, whose evasive species has, for half a billion years, matured in some primeval recess hidden from humankind. Out of the gloom, not one nautilus but two had materialized; they had mated before my very eyes and then vanished, leaving their secrets unrevealed and me alone with the sea's silent siren call: deeper, deeper. Down I had plunged by bathyscaphe, two thousand feet through the inverse universe, through the milky way of a billion glittering minute beings that implausibly multiplied rather than diminished in the cold and darkness three hundred fathoms down. Recently we had descended even farther; shortly before the nuclear conference, Doc Edgerton had joined my team to send a camera into the Romanche Trench, the equatorial canyon twenty-five thousand feet from the surface. To our astonishment we had found—five and one half miles from the light, the air, the land—a living creature, a "brittle star," aptly named. In the total darkness of the nether water world, beyond all expectation and against all odds, the lone brittle star's significance—its life—had shone like a sun.

All the while I had been delving into the sea, scientists had been delving into the atom. Its nucleus possessed hitherto inconceivable forces. That physicists had learned to liberate these furies during the course of a world war had led to the horrors of their being unleashed on civilian populations. With peacetime came the hopeful proposition that fission could be used not *against* human beings, but *for* them, as a source of clean, inexhaustible energy.

And so, with no preconceived concerns about atomic science—just great expectations about discoveries that lay ahead—I welcomed the IAEA scientists to our museum, turned their meeting over to them, and returned to my office.

The museum staff had installed an intercom system, and from time to time I slipped on a pair of headphones to listen to the IAEA proceedings. The scientists had gathered specifically to discuss radioactive waste. From the start, everyone agreed on the enormity of the problem presented by already swelling stores of irradiated debris. But no one agreed on how to sequester the toxic material for the thousands of years it would remain lethal.

One after another, technologists described their sometimes fantastical ideas: "fixing" radioactive waste in glass; burying it in salt mines; parachuting it onto the Greenland ice cap; trying to isolate it in deserts; rocketing it into space.

Up in my office, I listened uneasily as the scientists split into factions. The nationalists were easily identifiable; they indiscriminately vetoed all suggestions made by any of their countries' adversaries. The technologists, who had invested great effort in imagining futuristic proposals, seemed to have put almost no effort into considering the possible consequences of their plans. While physicists and chemists eagerly advocated trying some of the schemes, biologists—the life scientists—criticized the recklessness of proceeding with any option before identifying its repercussions, of setting off down uncharted paths that permitted no return. For their part, partisans of the nuclear industry—company spokesmen and government agents—interrupted and cut off all biologists who voiced misgivings. What an innocent I'd been, I thought glumly, in having thought that the museum could serve as host to a worldwide partnership of scientists and technologists.

Then suddenly I was an innocent no longer. Crackling over the headphones came the voices of a few technicians who urged that irradiated waste be cast into the sea. Their nations, they boasted, had even sunk some loads of radioactive waste already.

By the time I dashed back to the lecture hall, the room seemed to have ruptured. On one side, technologists, industrialists, and government agents urged full speed ahead, accusing biologists of being obstructionist; on the other side sat the biologists, stunned by their colleagues' infatuation with flashy technique, appalled by their indifference to danger.

I resolved not to allow myself to be swept along in the uproar. After all, I thought, one of the world's preeminent marine scientists was scheduled to appear the next day; certainly he would put an end to all discussion of ocean dumping.

When the illustrious expert arrived, I brought him to my home for lunch, where several of the distinguished international specialists, including officers of the IAEA, joined us. I had also invited Louis Fage, the marine biologist who had submitted my name for the museum directorship and whom I had come to regard as my teacher, my mentor, my inspiration, and my friend. We relaxed around the table and shared our thoughts, leaving behind the angry exchanges of the lecture hall. Someone mentioned ocean dumping. Our celebrated guest expert settled back comfortably in his chair.

"The sea," he began, "being obviously the natural receptacle for atomic wastes . . ."

Now my heart raced. I'd spent years diving among the oceans' astonishing life, fragile life, improbable life; but this great marine scientist had devoted his entire career to examining the most minute intricacies of the wonders I'd seen. How, I asked, could he even imagine allowing the seas to steep in radioactive poisons?

"Be realistic, Jacques," the man interjected. "There is only one problem ahead—the population problem. Soon we will have billions of people. We will have to feed them. We will have to develop nuclear energy without limit to run factories that will produce enough protein to feed humankind." He shrugged imperiously. "We must clearly accelerate the nuclear program even if disposal means eventually closing the seas to navigation." He looked my way and gave a dismissive wave of his hand. "Even closing the seas to all human activities."

I was stupefied. I glanced at my master, Professor Fage. I shall never forget his expression; his face was drawn in anguish. The others sat in hollow silence. The words droned on. I certainly understood the imperative of addressing population growth. But I couldn't see how we would mitigate a coming world problem by instigating a world catastrophe. I couldn't see how we would help teeming masses of people by closing off three quarters of the planet. I couldn't see why our technical ingenuity was not invested in finding a new, safe energy source rather than in contriving stopgap remedies for a hazardous one. But mostly, I didn't see the point of preparing a way for billions of people to inhabit a world in which there was no reverence for life.

Before that day, my career had been governed by my love for the sea. After that day, I devoted myself to defending the sea. I decided I could contribute most if I worked in cooperation with, rather than opposition to, atomic scientists; and so I offered laboratory facilities in the Oceanographic Museum to the IAEA, which at that time had not yet found a permanent marine-research base. We began a long and fruitful collaboration, during which I joined the scientists in selecting each year's research projects, most of which were connected to the measurement of ocean contamination from atmospheric tests and the effects of radiation on marine life.

Nearly forty years have passed. Nations passed the 1994 amendment to the London Convention, banning all ocean dumping of nuclear wastes. As for nuclear armaments, superpower stockpiles dwindled with the transformation of Eastern Europe, as the United States and former Soviet states dismantled thousands of their atomic arms. Back at my 1959 conference, even the optimists would not have predicted events as auspicious as these.

But another development has caught the world off guard as well; this one the experts have not failed to foresee so much as they have refused to acknowledge. After decades of enthusiastic production of nuclear energy and proliferation of nuclear weapons, we have glutted the planet with plutonium and highly enriched uranium. The Cold War is over. The Hot Peace has just begun.

The safe disposal methods that my conference atomists so confidently predicted have yet to materialize. Not one single nation has put a permanent method of waste storage into operation; containers in which refuse is packed swell with radioactive gas and leak into the open dirt. Hundreds of aging nuclear power plants approach obsolescence while the companies and countries that built them have neither the money to dismantle them nor the place to bury the contaminated parts. Government agents have sidestepped the problem of safeguarding people by ignoring it, draining radioactive liquids into rivers and pumping tons into the earth, using the stamp of "national security" to keep their dirty secrets hidden from the eyes of their own people.

Weapons negotiators have had no more success than waste guardians in gaining control over the atomic demons flying from Pandora's jar. Nuclear nations insist that others forgo stockpiling of atomic weapons even as they continue to augment their own. Self-righteous when preaching nonproliferation, the nuclear nations suspend their sermon at the mention of profits, ever ready to peddle their atomic hard- and software to any warrior state that can afford to pay. Countries at the customer's borders see their status change from neighbor to potential target and swear to acquire their own means for atomic revenge. Into this volatile situation now step thieves and smugglers, hawking stolen uranium and plutonium in the marketplace of tyrants and terrorists. The end of the Cold War may have fulfilled our wildest dreams, but the beginning of the Hot Peace realizes our worst nightmares.

In the meantime, those in charge reassure rather than resolve.

Nuclear-energy executives promise that plutonium from power plants poses no weapons risks, even though they know that some nations have already diverted plutonium from energy facilities to bomb facilities. Government officials preach that atom-bomb use is unthinkable even as they contend that atom-bomb building is mandatory. Armaments negotiators congratulate themselves for the fact that no more than five nations have "declared" themselves as nuclear powers—as though, by refusing to acknowledge all the *undeclared* nuclear powers, they will somehow make the hidden stockpiles disappear. Our leaders assert that they cannot ban plutonium and highly enriched uranium and must resign themselves merely to regulating them. We still await the day when sane people everywhere read the clear, plain print of history and rise up to contend that fissile materials have *never* been effectively regulated; that they *cannot* be effectively regulated; and that therefore they *must* be banned.

As our proliferation potentates fiddle, millions dwell in cities that can be devastated by a terrorist's crude atomic weapon—delivered not on an incoming missile but in an innocuous suitcase. Seven hundred million of our world's people live in homes less than one hundred miles from an aging power reactor that could disgorge its radioactive materials in an accident. When I have voiced my protests over nuclear issues, I have often received messages from technicians and even politicians reminding me that I should stick to steering boats, that proliferation is not my business. They're right. It's everyone's business.

* * *

The cold facts of today point to the urgent necessity of facing tomorrow. Yet looking backward is important as well; its value lies not in fault-finding but in path-finding. If we can understand what led humanity into the atomic predicament, we might learn how to lead humanity out. We can see that precedents set in wartime have forever after disfigured the peace.

Those who blame our atomic woes on atomic scientists head down the wrong path at the outset. When Henri Becquerel innocently discovered radioactivity in 1896 and Albert Einstein recognized the relation between matter and energy in 1905, their sights were trained on the world of learning and discovery, not the world of war. Einstein himself protested that he could not have envisioned the consequences of his equation. "You suggest," Einstein would later write to an accuser, "that I should then have foreseen the possible development of atomic bombs. But this was quite impossible since the accomplishment of a 'chain reaction' was dependent on the existence of . . . data that could hardly have been anticipated."

How could Frédéric and Irène Joliot-Curie have felt any emotion other than astonishment when they succeeded where centuries of sorcerers had failed—when they performed veritable atomic alchemy? The two scientists altered the immutable: By bombarding an element with neutrons and knocking a few protons from the nuclei of its atoms, they had actually changed one element into another nearby on the periodic table. An enthralled Enrico Fermi set off bombarding all the elements. When Fermi bombarded uranium, however, wonder turned to bewilderment. Fermi transformed uranium into another element, to be sure—but one, he thought, unknown to humankind.

In Germany, Otto Hahn and his colleague Fritz Strassmann focused on identifying this unidentified element. They too bombarded uranium and found, to their astonishment, that the resulting element was not at all unknown; it was barium. They were stunned. If they had changed uranium atoms into barium, their nuclei had lost not just a few protons, but almost half. Hahn and Strassmann had not merely changed uranium into an element nearby on the periodic table. They had changed it into one that was not even nearly related.

Physicists who had been following their work instantly guessed that they had not simply bumped protons from a nucleus; they had actually split one nucleus in two. They had achieved fission of the very heart of matter, releasing unimaginable power. The two splitting halves of a nucleus, physicists calculated, would fly apart with a total kinetic energy of nearly two hundred million electron volts. When this news was communicated to Niels Bohr, the great nuclear physicist reportedly exclaimed: "Oh, what idiots we all have been! Oh, but this is wonderful! This is just as it must be!"

That those involved were blind to the direction in which the world was heading is underscored by the fact that the Nobel Prize—awarded to those who "have conferred the greatest benefit on mankind"—was awarded for each one of these discoveries. Time inflicts painful hindsights: The unwitting forefathers of the atomic bomb received the honor created by the inventor of dynamite. Alfred Nobel, known in Sweden as a "mad scientist" because his explosives blew up his factory along with his youngest brother, died a pacifist who hoped the destructive power of his invention would bring an end to war. Tragically, the Nobel Prize that has been most frequently withheld—due to lack of a worthy candidate—is the Peace Prize.

At this time, the glittering events taking place within the laboratory were shadowed by the gathering storm outside. One man who feared the implications of all this atomic wizardry's being performed during the course of a world war was the thermodynamics specialist Leo Szilard. Having fled wartime Hungary and finally finding refuge in the United States, Szilard considered the diabolical possibilities—that the power unleashed by splitting atoms of uranium could be used in a bomb a million times more powerful, per pound, than any known explosive, and that Hitler might build such a weapon before others could. Szilard later described the evening in 1939 during which he duplicated the fission experiment: If flashes appeared on a cathodic tube, he would have confirmed that the energy of the atom had indeed been liberated. "I turned the switch . . . saw the flashes . . . watched for about ten minutes; and then switched everything off and went home. That night I knew the world was headed for sorrow."

Roused by the news that an atom had been split, Szilard feared that the nightmare power of a "chain reaction"—in which one splitting atom set off a sequence of splitting atoms—would soon be held in human hands. He contacted other physicists, beseeching them to keep their ongoing findings in the "dangerous zone" to themselves. Szilard promptly collided with a sad truth about human nature, and about nationalistic pride, that has gone on to define the atomic age: Few can resist the allure of sovereignty, primacy, and prestige accorded to nations that rule the nuclear arena. In France, Frédéric Joliot-Curie succeeded in setting off a chain reaction and was promptly seduced. With dreams of generating limitless electricity and thereby rendering our country industrially supreme, Joliot-Curie ignored Szilard's pleas for censorship. He openly published the precise measurements he had used to ignite his chain reaction in a mass of uranium. He thereby set off a chain reaction of another sort. He set off the arms race.

In short order, American, British, and Soviet scientists supplied their governments with information regarding the possibility that fission of uranium atoms could be exploited in a weapon. The world's greatest supply of uranium, at the time, was being mined in the Belgian Congo. Britain arranged to earmark significant quantities of the Belgian uranium stores. Before long, Holland too had ordered fifty tons.

Then came ominous news: Shortly after Joliot-Curie in France had published his measurements for a chain reaction, Germany prohibited sales of uranium from the mines it had seized in Czechoslovakia. How far had the Germans progressed in their own nuclear research? Hitler's troops were heading toward Belgium. What would happen if the Nazis, already wielding absolute control over uranium in Czechoslovakia, took over the large quantities of uranium in the Belgian Congo as well?

Szilard, aware that his friend Einstein knew the queen of Belgium, asked the great scientist to warn her. Up until that time, Einstein had resolutely remained an innocent, preoccupied with his theoretical musings. Wartime jarred him awake. An ardent lifelong pacifist, Einstein had fled Germany after having been denounced as a Jew and having had a price put on his head for his murder. The Nazi threat was all too real to him. At Szilard's urging, he not only consented to warn the Belgians about the dangers of German control of uranium but also made a decision that he would later describe as "the one great mistake in my life." Einstein agreed to sign a letter, which Szilard would draft, urging U.S. president Franklin Roosevelt to obtain a supply of uranium ore and accelerate nuclear research on the side of the Allies.

In the letter, Einstein and Szilard described chain reactions and noted that "it is conceivable that extremely powerful bombs of a new type may thus be constructed. A single bomb of this type . . . might very well destroy the whole port together with some of the surrounding territory." Roosevelt, who also received warnings from British scientists, gave his unequivocal approval for the Manhattan District Project, in which American and British physicists would join in a full-speed effort to build the bomb.

The two precedents set thus far—first, racing an enemy to conceive an atomic weapon, and second, dreaming of national prestige through nuclear energy—can be understood in the context of a world at war, in which national identity and even national survival were at stake. The third precedent, however, was less a response to the atrocities of a Hitler than a surrender to the seductions of technology. Throughout the world, scientists once motivated by high ideals underwent a kind of dark alchemy of their own, turning into technocrats too often titillated by the wonders and powers of the atom.

It was a German soldier who gave me my first glimpse of the black-magic allures of the new findings. During the war, near the close, when the future was bright for France and dark for Germany, I was serving as an intelligence officer collecting information on Nazi and Fascist forces. One day an agent of mine told me a German officer had unwittingly confided that, although the situation didn't look promising for the moment, the Nazis would soon have a surprise to spring on the world: "We have *la poudre de Perlimpinpin*." In an age-old French fairy tale, Perlimpinpin the magician sprinkled magic powder to make things and people vanish into thin air. I wondered what kind of new weapon the German soldier was describing. It was only in retrospect—after I'd read descriptions of human beings reduced to shadowed silhouettes in concrete at Hiroshima's ground zero—that I felt his charming allusion's chilling implications.

Anyone, from any country or alliance, could lose his soul to Perlimpinpin. Hundreds of U.S. scientists were seduced into joining the Manhattan Project during the war by what one admirer called the "intellectual sex appeal" of Dr. J. Robert Oppenheimer, the project's director and energetic recruiter. Hitler's subsequent suicide and Germany's defeat seemed to leave Oppenheimer's team feeling somewhat deflated. Deprived of what was supposed to be their motivation, the technologists discovered they could nonetheless vent their talents on something—if not on beating Hitler to the bomb, then perhaps on contributing to a decisive victory. Oppenheimer himself clearly voiced his enthusiasm at having found a raison d'être for his project even after Germany's defeat: "We were still more frantic to have the job done so that if needed, it would be available.

"Almost everyone knew that this job, if it were achieved, would be a part of history," Oppenheimer wrote. "This sense of excitement, of devotion, and of patriotism in the end prevailed . . . I have never known a group more . . . devoted to a common purpose . . . more understanding of the role that they were playing in their country's history."

Having succeeded in reducing scientists to servants, military personnel took another step in advancing their authority. They instituted a super secrecy around the scientific truths and principles that were being applied to create weapons. They withheld information not only from the enemy but even from decision makers and from important government advisers. Anyone with a warning, a reservation, or a suggestion in the geopolitical, philosophical, or moral sphere was sealed out of decisions to build a weapon that would change the course of civilization. Human beings were left out of the human equation.

Illustrations include not only the omission of pertinent information needed by decision makers—few were told much about radiation sickness, for example—but also the exclusion of knowledgeable experts who could have informed the decision-making process. Many of the Manhattan Project's own scientists signed a memorandum drafted by James Franck, the only scientist who had premised his participation on the condition that he would have the opportunity to inform government officials of his views. Franck urged officials to try to avert an almost certain postwar arms race by demonstrating the bomb "on a desert or barren island" as a warning before dropping it on human beings. "Statesmen who did not realize that the atom had changed the world," Franck wrote, "were laying futile plans for peace while scientists who knew the facts stood helplessly by."

The scientists who signed Franck's memo then stood helplessly by while their report was submerged in a review committee. Leo Szilard, also troubled by the fact that scientists working on the nuclear bomb were barred from knowing the military's nuclear plans, took his own concerns to U.S. secretary of state James Byrnes, who responded by expressing his disgust that a mere physicist should "desire to participate in policy making" regarding his own discoveries. The impassioned Szilard was further targeted by General Leslie R. Groves, who called the scientist "a parasite living on the brains of others" and even went so far as to try to get him interned during the war.

The atom bomb was dropped on Hiroshima August 6, 1945. The copilot of the American bomber, Robert Lewis, would later recall his shock at seeing the city one moment and the next moment, seeing nothing—nothing but a rising column of radioactive ash. The plane heaved in the atomic winds, then swerved to return to its base. "My God," Lewis noted in his journal. "What have we done?"

Some hundred thousand people died in the blast.Three days later, another nuclear bomb killed tens of thousands more in Nagasaki. Radiation sickness in the ensuing days, months, and years took more than another hundred thousand lives.

President Harry Truman publicly noted "the tragic significance of the atomic bomb" and yet thanked God that the "awful responsibility" for it had "come to us, instead of to our enemies . . . We must constitute ourselves trustees of this new force." In proclaiming his assurance that the exclusive group of countries that already possessed bombs could unilaterally proclaim themselves the ongoing decision makers of the international nuclear arena, Truman unwittingly articulated the last nuclear precedent set in wartime. He institutionalized the caste system of the atomic age, with its haves and have-nots, founded on the notion that those who brandish nuclear weapons are empowered to live in security and superiority, and those who do not are condemned to live in fear and inferiority.

Now the two nations that had joined forces to create the bomb—Britain and the United States—joined forces to contain it. Leading citizens who had been excluded from bomb decisions during the war drafted a proposal for the United Nations to set up the Atomic Development Authority. They envisioned an agency that would control all aspects of the nuclear process—govern all uranium sources, all reactors and separating plants to produce and enrich fissionable materials. It would be staffed by scientists from all nations. Diversions of fissionable materials would be readily detected and immediately punished.

The Soviets, who had been secretly conducting a uranium project that Stalin had authorized at the peak of the war, some three years before, rejected every proposal. The Americans, guarding the secret of how much further their bomb program had progressed, refused to negotiate.

The physicists now stepped forward to try to place decisions about the bomb in the hands of those whom it endangered: the people. Who better than the scientists to communicate that if they had been able to follow the nuclear path from its promising start to its odious finish line, others could too? The censorship of scientific papers meant nothing; fission, chain reactions, atomic forces—all were written across the universe, inscribed within the atom.

Einstein chaired the Emergency Committee of Atomic Scientists; Szilard cofounded another group called the Federation of Atomic Scientists. Both had the declared purpose of taking their knowledge directly to the public. Eventually Szilard even approached Nikita Khrushchev and persuaded him to install the famous direct line between Moscow and Washington. Joliot-Curie founded the world peace movement and issued the Stockholm Appeal. Oppenheimer, who publicly expressed contrition and compellingly condemned the military's denial of bomb-related information to decision makers, rallied for public education, organizing Operation Candor. Their messages underscored the same points: Atom bombs will become more destructive. There is no military defense. Preparedness against atomic war is futile and if attempted will ruin the structure of our social and economic order. The solution must be invested with an authority equal to the threat.

The military reacted with the same kind of contempt it had reserved for freethinkers during the war. When Oppenheimer opposed the U.S. hydrogen bomb project, he was officially accused of subversion, and his security clearance was revoked. The United States tested its first hydrogen bomb in 1952. Joliot-Curie, having founded his country's nuclear program in the name of peace, was fired when the time came to produce bombs. The Soviet Union joined the other countries in this kind of suppression, with Andrei Sakharov "seized by a feeling of impotence and fright" when he realized he had no authority over the hydrogen bomb he had created.

And so, with little public participation allowed and no expert dissent tolerated, the "peacetime" atomic arms race began. U.S. general Leslie Groves, dismissing scientists' warnings, estimated that the Soviets could not assemble a bomb for twenty years. He urged Truman to launch unmitigated American bomb production and refinement. Truman, who concurred, remarked in 1953 that he was "not convinced that the Russians have achieved the know-how to put the complicated mechanism together."

- Britain, in almost a complete absence of public debate, exploded its first atomic bomb on October 3, 1952.

- The Soviets exploded their first test of a true hydrogen bomb in 1955.

- De Gaulle formed the Atomic Energy Commission in France two months after the bombing of Hiroshima. France exploded its first test bomb February 13, 1960.

- The Chinese exploded their first test bomb October 16, 1964.

- India, feeling threatened by China, exploded a test bomb in 1974 and may have obtained the materials to build almost two dozen nuclear weapons and to develop hydrogen bombs.

- After India exploded its first nuclear bomb, Pakistan launched a covert bomb program.

- Israel has been reported to have secretly stocked enough plutonium for anywhere from fifty to two hundred warheads. The country is said to have the capacity to make thermonuclear weapons.

○ Japan—which, mourning its incinerated cities, officially decried the American use of an "inhuman weapon" in 1945—had in fact been working on an atom bomb project of its own beforehand and, in the words of one participating physicist, "had no doubts about using it if we could." While voicing its distrust of neighbor North Korea, Japan today is said to have become a "virtual nuclear power," possessing all the necessary elements as well as the capacity to create a bomb within weeks.

○ North Korea, which accuses Japan of using its civilian energy program to "race headlong toward nuclear armament," has itself been suspected of working on weapons. Iraq was alleged to have long run a covert nuclear-weapons program before the 1990 Gulf War. Algeria's nuclear-energy program is suspected of concealing a secret weapons project, and Libya also poses a nuclear-weapons risk. Iran has purchased equipment useful for atom-bomb building and is expected to achieve its own bomb in a short time.

○ Brazil, Argentina, and South Africa announced that they were abandoning their bomb projects before the international disarmament bureaucracy would even acknowledge they had them. Experts warn that the bomb-making abilities remain.

○ Sweden, Germany, Italy, Switzerland, Canada, South Korea, and Taiwan are all reputed to possess both the expertise and the material to manufacture bombs. Australia, Austria, Belgium, Denmark, Egypt, Finland, the Netherlands, Norway, Spain, and Yugoslavia possess the necessities for building bombs in a relatively short time.

* * *

"The atomic bomb is too dangerous to be loose in a lawless world," Truman said in his radio interview after Hiroshima. What happened? Certainly today's proliferation of weapons is partly the product of the wartime precedents. Foremost among those trends is the notion that nuclear capability bestows national leverage, influence, and prestige. Immediately after World War II, membership in the nuclear club seemed as enviable as it was exclusive, with wealthy and developed nations boasting of their bombs and impoverished, nondeveloped nations wishing for them.

My own country surely waded into the atomic quagmire in quest of grandeur. The lure of national prestige that had tantalized Joliot-Curie also captivated General Charles de Gaulle. As detailed by McGeorge Bundy in his accomplished history, *Danger and Survival: Choices About the Bomb in the First Fifty Years* (to which we are considerably indebted), in 1954 the United States passed a major amendment to its Atomic Energy Act of 1946, forbidding the exchange of any nuclear secret with any foreign nation, but shortly thereafter exempted England from its blackout while leaving France in the dark. De Gaulle wrote the U.S. president Eisenhower to express regret that France had been literally left to its own defenses; Bundy quotes de Gaulle's aide, the prominent diplomat Geoffroy de Courcel, as explaining that for de Gaulle "what mattered was to be able to say that France was politically a nuclear power, and to talk as equal to equal with the Anglo Saxon powers." That it was primarily offended pride that triggered France's nuclear armamentation seems even more clear in de Gaulle's telegram to the country after our first bomb test in 1960: "HURRAH FOR FRANCE! SINCE THIS MORNING SHE IS STRONGER AND PROUDER."

That same year, I received word that President de Gaulle planned to visit our Oceanographic Museum in Monaco. At the time, not only was I serving as director of the museum, but I was also protesting the French government's proposal to dump nuclear waste into the Mediterranean. Shortly before de Gaulle was scheduled to arrive, Prince Rainier of Monaco received a telephone call from the Élysée Palace; a functionary told Rainier to bar me from the museum the day of the presidential tour. With some amusement, Rainier told me about the conversation and, of course, declined to put pressure on me.

I decided to receive the president personally. As his limousine arrived, I stood, outfitted in my full Navy uniform, at the top of the steps ascending to the museum's doors. When he stepped from the car, I did not budge; I waited for him to climb the staircase. Then I declared, "Mr. President, welcome to the temple of the sea!" I guided him through the museum and then took him to the back windows, which overlook a cliff dropping to the Mediterranean. Directly below, my three ships, *Calypso*, *Winnaretta Singer*, and *Espadon*, dressed overall, flags whipping in the wind, bounded on stormy waters. I announced, "My ships have come to greet you!"

I intended to communicate not only respect but also resolve. De Gaulle was undiverted. As soon as he was alone with me in a tiny elevator, on our way to the next floor, the president commented dryly, "Cousteau, be nice to my atomists." I replied, "It is your atomists who should be nice to us." By the time we reached an upper floor, newspaper photographers and even political cartoonists were having a field day, recording their images of de Gaulle and me as we glared at one another and as Rainier and Princess Grace, at the end of the room, politely cast their eyes downward and looked as though they would rather be anywhere else on earth. I tried to block the circus out of my mind and pressed on, telling the president what I believed: I regretted that nothing more than the U.S. decision to share its nuclear secrets with England and not with France had compelled him to turn our country into a nuclear fortress. We were not defending ourselves; we were defending our honor.

Others, inside the borders of France as well as out, have continued to share de Gaulle's now-anachronistic notion that the atomic bomb brings prestige. When much of the rest of the world expressed dismay at India's detonation of a nuclear explosive, the chairman of France's Atomic Energy Commission responded by sending India a congratulatory telegram. Even now, in a world in which the use of an atom bomb would be considered the act of a maniac, China's national-nuclear-corporation chief has rhapsodized that "nuclear industry and technology are a yardstick for measuring the overall strength of a country. [They] provide a pillar for its international standing." North Koreans nurse the same delusion: When a foreign journalist accomplished the "impossible" feat of penetrating the country and living among its people, he saw workers with rags wrapped around their feet in lieu of shoes and heard anguished tales of deaths from malnutrition. But when he asked North Koreans about the nuclear program of Kim Il Sung, he found that "most people are satisfied that they endured near starvation in order for the government to develop a weapon that brought the rest of the world to its knees."

*   *   *

The precedents set by war, however—the fear that an enemy will get the bomb first; the supposed prestige that comes with stockpiling bombs— are not alone in fanning the flames of proliferation. Today the spread of bomb material is largely fueled by the hypocrisy of the very nations who say they seek to control it.

Signatories of the Nuclear Nonproliferation Treaty (NPT), regarded as the cornerstone of disarmament efforts, have in particular perpetuated and promoted the worldwide spread of fissionable materials.

Most critics complain that the NPT document itself is unfair; they describe its provisions as biased against the developing world. Many of the nonnuclear nations that signed the treaty have later regretted their decision to do so, contending that they were betrayed into accepting a situation in which they would be cast as helpless supplicants, pleading for atomic protection and advanced technology from the five countries that anointed themselves as permanent atomic powers. Nonnuclear nations resent the fact that the nuclear powers developed atomic energy without surveillance but now force supervision on them. They condemn the NPT, as do most commentators on arms control, as a document that is discriminatory, favoring the haves at the expense of the have-nots.

But it is not the Nuclear Nonproliferation Treaty itself that has created and perpetuated the gap between the nuclear haves and have-nots. The nations who break the treaty have done so. Since its inception in 1968, the NPT has steadfastly claimed one goal: eliminating nuclear arms. The treaty ruled that the arms race would stop in its tracks at its 1968 status, then shift into reverse until all bombs were gone. Those with atomic bombs would begin negotiations to dismantle them; those without atomic bombs would swear never to build them. The IAEA, as an independent U.N. department, would act as the treaty's "watchdog."

With what dedication have nuclear nations honored their sworn commitment to achieve the "cessation of the nuclear arms race . . . and . . . complete disarmament under strict and effective international control"? After the Soviet Union and the United States signed the treaty, the two countries more than doubled the number of nuclear bombs in their silos. Although U.S. officials called on other nations to extend the NPT "indefinitely and unconditionally" when it came up for its 1995 renewal, the U.S. administration, for its part, published its own official stance on stocking, and using, strategic weapons. More than a quarter century after signing the NPT, the United States emphasized the continuing role of nuclear weapons in U.S. strategy and refused to rule out nuclear retaliation against even a nonnuclear attack. Even if all the celebrated post–Cold War disarmament pacts are honored, the two countries will only return their stores of long-range nuclear weapons to the levels they stocked at the birth of the treaty. The twenty-first century will open on a planet where nations aim some twenty thousand nuclear warheads at each other.

The nuclear *Club des Grands* has also shown contempt for another NPT provision, which requires that bomb nations discontinue "all test explosions of nuclear weapons for all time." With a comprehensive test ban, existing weapons would gradually become unreliable; nuclear powers would gradually lose confidence that their bombs would explode if used; and arsenals would become, at last, obsolete.

Nuclear powers spent twenty-five years concocting delays and excuses for disobeying this provision. Then came a moment when the world's people had reason to hope. In 1991 the U.S.S.R. announced that it would discontinue its test explosions. France declared a moratorium on its test explosions in 1992; the United States and Britain followed suit. All signs indicated that nations would formally sign a comprehensive test-ban treaty being drafted in Geneva. In May 1995 came another auspicious sign: Just as the NPT reached its expiration date, nations agreed to extend it indefinitely.

Days later, China exploded a test bomb. Weeks after that, Jacques

Chirac launched his new presidency of France with the flat declaration that he would break the moratorium on testing and explode up to eight more bombs.

All the declared nuclear nations together, including even China and France, attempted to mute public reaction to renewed testing with more promises to sign a comprehensive test ban the following year. But to what extent do leaders who make such pledges and even sign such bans bank on the public's gullibility, its short-term memory? The United States did indeed stop exploding test bombs—but only when it could test newly invented nuclear weapons without explosions, by using computerized simulation technology in the laboratory. The panel of military experts that urged the French government to detonate more bombs openly asserted that the tests were needed in order to upgrade computer technology so that France, too, could modernize nuclear weapons without explosions after the test ban. According to Pascal Boniface, the director of the Institute for International and Strategic Relations in Paris, France hopes to create a low-radiation neutron bomb as well. China, also, has been suspected of conducting its tests in order to create new weapons: The head of Stockholm's International Peace Research Institute contends that China is developing new, miniaturized warheads that, when loaded on a single missile, will be capable of accurately striking multiple targets.

Such developments indicate that not only is the public's safety being compromised; its intelligence is being insulted. The stated goal of the NPT, and of the comprehensive test ban it requires, is not to eliminate underground tests; it is to eliminate *all* tests—and by so doing, all bombs. Nations that hurry to perfect new weapons before signing a treaty that prohibits other nations from doing so are motivated not by altruism but by arrogance. A ban that prohibits only obsolete underground tests but permits advanced laboratory tests is not a "comprehensive test ban." It is a counterfeit test ban.

Given such hypocrisy, it would appear that nuclear powers assume themselves to be above the rules, which would indeed justify the accusations of unfairness lodged by the nuclear have-nots—were it not for the fact that they break the rules just as brazenly.

Their violations involve exploiting and perverting the treaty's provisions regarding using the atom to generate nuclear energy, thought at the time the treaty was written to be the key to progress and prosperity. The treaty did not require its nonnuclear signatories to forswear this bounty. A nonnuclear country had only to sign the NPT to purchase fissionable material at moderate prices from any of its fellow nuclear signatories. In return, the nonweapons nations would promise not to transfer or receive atomic-arms-related technology, not to divert fissionable materials from their new power plants to any secret weapons works, and finally, to verify their honesty by opening their nuclear programs to inspectors from the IAEA, which would periodically inventory the plutonium produced. If inspectors discovered that plutonium was missing from the power plants—presumably diverted to a secret weapons project—the other signatories would join to impose sanctions on the illicit bomb builder.

Both North Korea and Iran have demonstrated their indifference to the stipulations of the treaty they signed. Just four years after North Korea signed the NPT, the country was suspected of diverting nuclear fuel to a weapons facility. North Korea bought time simply by denying inspectors access to its operations. During ensuing negotiations with an international consortium that begged North Korea to freeze its nuclear program, the country continued to deny having bombs while at the same time repeatedly issuing threats hinting that it not only stocked bombs but also would use them. North Korea furthermore suggested that if it did not receive economic and political payoffs in exchange for freezing its nuclear program, workers would begin reprocessing the country's spent atomic-power fuel, from which they could separate out enough plutonium for multiple warheads.

Iran too has disregarded the obligations it accepted by signing the NPT. Western officials long monitored Iranians' "intensive effort to get everything they need" for a nuclear bomb, which, according to reports published in 1994, the country was expected to achieve in anywhere from ten to "only a small number of years" hence. Fifteen years beforehand, Iran had ordered nuclear-energy reactors but had never received them, after abandoning the shah's atomic-energy program during its revolution. In 1994, Iran renewed requests for delivery. Wary of Iran's bomb ambitions, several nuclear nations sought to block delivery of the reactors by asking the merchants—Germany, Russia, and China, among others—to refuse shipment. An indignant Iran maintained that, as an NPT signatory in good standing, it had a right to nuclear goods; after all, inspectors from the IAEA had been unable to locate hidden weapons works. For their part, declared nuclear powers complained that the NPT was so weak that Iran could even develop bombs while remaining in "strict compliance" with the feeble treaty.

The average citizen, of average intellect, is well enough equipped to ask the obvious question: What goes on here? A country that is secretly building bombs is not "in strict compliance" with a treaty that forbids them; nor is it an NPT member "in good standing." In order to receive the NPT's benefits, a nation should not only sign the treaty; it should also obey the treaty. Nonnuclear nations that complain of discrimination when they cannot buy bomb materials, merchant nations that complain of the NPT's inadequacy even as they pocket profits from dangerous sales—they all have mutilated the NPT's purpose. That purpose is not equality in a world in which *all* nations stock bombs; it is equality in a world in which *no* nation stocks bombs.

* * *

The international accusations, counteraccusations, and righteous indignation serve only to obfuscate the real reasons why the Nuclear Nonproliferation Treaty has proved ineffective. That explanation lies not in the allegation that its rules are unfair, but in the fact that its rules are not enforced. The countries that have signed the NPT refuse to invest it with authority; they deny power to the very agency they themselves have designated to control the most formidable power on the planet.

NPT member nations tell the world that the IAEA acts as their treaty's "watchdog," but the agency is a watchdog that they themselves have tried to render both toothless and blind. NPT signatories that are declared nuclear-bomb powers deny IAEA inspectors entry into their nuclear complexes. The United States has even stated outright that it prohibits IAEA inspections because it does not want to foreclose military options to reuse plutonium from decommissioned nuclear weapons, the very option it supposedly relinquished by signing the NPT.

As for NPT nations that have forsworn nuclear-bomb projects in the supposed trade-off for nuclear energy: They allow IAEA agents to enter only atomic facilities that they have openly declared to be energy plants. Thus agents can testify that honorable nations behave honorably. But as for those without honor: They refuse the IAEA the authority to search for hidden caches of fissionable materials on plant property. Agents have no recourse when nations delay or cancel inspections. If by luck investigators happen to find a weapons works, they have no permission to enter without first winning near-unanimous approval from the U.N. Security Council. Even with such dispensation, IAEA inspectors are trained only in the commercial power cycle; they have no expertise in nuclear-bomb projects and so are ill-equipped to investigate covert weapons programs.

This emasculation of the NPT's enforcement authority points to one conclusion: The primary aim of NPT nations has never been to protect the world's peoples; it has been to protect the nuclear marketplace.

The accusation seems harsh; it seems cynical; but, to the average citizen, it also seems irrefutable, given the charter documents. The Nuclear Nonproliferation Treaty guarantees to all signatories the "inalienable right" to purchase nuclear-energy matériel at moderate prices. The treaty promises both a world without weapons and a robust world nuclear-energy market. No one who speaks the truth speaks of a world that can have both.

* * *

The liaison between energy and bombs remains as obvious today as it was during World War II, when alarmed physicists warned their governments about the common element shared by atomic bombs and atomic power: uranium. Scientists' concerns centered specifically on one particular group of isotopes within the ore—uranium 235, the only uranium isotope in which atoms split.

U235 isotopes, extremely sparse within ore, constitute just one tenth of 1 percent of natural uranium. In order to exploit U235's fearsome fission powers, bomb builders needed a concentrate of the isotope. They learned to "separate" it out of ore and then to create a "highly enriched," weapons-grade uranium, a walloping 90 percent of which consists of U235. By midwar, physicists found an even easier way to produce fissionable material. They discovered that when they simply ran uranium through a reactor, the resulting mass of irradiated metal contained isotopes of an entirely new element that does not occur in nature: plutonium. They also learned to "separate" plutonium isotopes from the irradiated metal mass. The weapons-grade plutonium they produced is even more fissionable than highly enriched uranium, with far less required for a bomb.

The catch that unravels the fabric of present nonproliferation treaties: When atomic-energy plant employees run uranium through their reactors, splitting atoms to release power, they too, of course, find that the resulting irradiated mass—now called "spent fuel"—contains both U235 and plutonium. They too can separate both these fissionable materials out and concentrate them, to a so-called reactor-grade level, which they can then reuse as new fuel—meaning that the highly fissionable bomb-building materials plutonium and U235 are present throughout the world at civilian energy plants.

Repeatedly and appallingly, those with ties to the nuclear-energy market mulishly persist in misleading the public about atomic power's relevance to weapons proliferation. Officials at France's state-owned reprocessing company Cogema have openly belittled the French people for confusing weapons-grade plutonium with the kind of plutonium that is produced in civilian reactors—shamelessly providing the misinformation that reactor-grade plutonium is unsuitable for use in bombs. The director of atomic energy policy at Japan's Science and Technology Agency likewise has remarked that the "Western press," in publicizing Japan's mountainous plutonium stores, consequently overestimates his country's nuclear-bomb capacity. He had the audacity to declare that "at present there is no technology in the world to refine reactor-grade plutonium into weapons-grade plutonium."

Yet the press is not responsible for the idea that nuclear-energy reactors produce plutonium usable for bombs. Physics is. Plutonium, by definition and discovery, is produced by irradiating uranium in a reactor, and nuclear-energy reactors are no exception. *All* forms of plutonium can, in turn, be used to create a nuclear explosive. Reactor-grade plutonium from a nuclear-energy plant can be made into a weapon that, though experts designate its power as "primitive," would nonetheless be capable of a knockout blow equivalent to more than a thousand tons of high explosive. By contrast, the nonnuclear explosives that in 1995 blasted the Oklahoma skyscraper weighed but two short U.S. tons and did not leave behind the lethal radiation legacy that would have taken tens of thousands more lives. Worse, even after plutonium is processed into reactor-grade fuel, it remains in readily separable form and need only be reprocessed to be purified into weapons-grade plutonium fearsome enough for national strategic arms and intercontinental annihilation. The U.S. National Academy of Sciences has stated it starkly: "Reactor grade" plutonium is a "clear and present danger to national and international security."

The suggestion made by energy-industry officials—that the public is irrational in its concerns about nuclear energy's connection to nuclear bombs—is not only manipulative but also clearly inconsistent with current events. If, as officials insist, atomic energy produces no materials for atomic weapons, why did disarmament negotiators and intelligence operatives raise an international furor over North Korea's atomic-energy program? Why did American nuclear experts converge on North Korea to supervise removal of its spent nuclear-energy reactor fuel rods, laden, they said, with bomb-potential plutonium? Why did Iran's request for nuclear-energy reactors compel intelligence operatives to sound a worldwide alarm?

Finally, the contention that civilian nuclear-energy plants feed nuclear proliferation is not mere theory; it is historical fact. India made its bomb from plutonium produced by a "peaceful" nuclear reactor, purchased ostensibly for its civilian program from Canada. The Soviets used civilian power reactors for military purposes. A host of other countries, including Japan, Germany, and Algeria, are judged to have come closer to bomb-building capability through rigorous civilian nuclear-energy programs.

A bomb builder needs just fifty-five pounds of highly enriched uranium to fashion an explosive, or just seventeen pounds of plutonium. Plutonium, furthermore, is relatively easy to smuggle—its alpha rays do not penetrate far, eliminating the need for thick lead shielding—and it can be used for genocide without even the need for a chain reaction. One of the most lethal poisons on the planet, a mere grain of plutonium—one thirty-thousandth of an ounce—will cause cancer if inhaled; millions of people could perish if just ten ounces were sprinkled into a city's water supply or simply burned outdoors.

And just how much plutonium and highly enriched uranium have we human beings produced? Over how many pounds of fissionable materials must we—and our children, and their children—remain vigilant? The IAEA estimates that in the past fifty years, militaries and nuclear-energy industries have disgorged about a thousand metric tons of plutonium. Much more than half of that plutonium has been produced solely by nuclear-energy programs, which have already separated out 120 tons of plutonium from their spent waste and processed it into a form useful either for recycled fuel or for weapons. Yet no one on Earth, not even the NPT's watchdog the IAEA, can testify for certain to just how much plutonium exists. Individual governments and corporations refuse to release their inventory statistics, claiming military and commercial confidentiality.

No one has more right to data concerning the world's most lethal substances than the world's citizens. And any assembly of nations genuinely committed to defending the people from proliferation would demand proliferation data. Perhaps the real reason for suppressing the figures has less to do with protecting confidentiality than it does with protecting the open secret of the atomic age: Nations that sign a "nuclear nonproliferation" treaty but that sell nuclear-energy reactors, sell uranium, sell plutonium, are selling bombs. Selling nuclear energy around the planet is selling the security of the people who inhabit it.

*   *   *

Is nuclear energy worth it? Is it that cheap? That profitable? That clean? In a perfect world, a technology would be valued according to its benefits to human health and safety. But even in an imperfect world, where value is counted in coins, is nuclear power a wise financial investment?

In many countries, such as my own, where nuclear energy remains largely under government control, the financial data to answer that question have been kept from the public that must pay to balance the books. On the other hand, although nuclear energy was conceived with some secrecy in the United States, it was ultimately born into the open market—where the relevant figures are both available and persuasive. Nuclear power has proved to be a financial fiasco.

After Hiroshima, the American phrase that resounded throughout the world—"Atoms for Peace"—inspired hope; it seemed that human beings finally would apply their intelligence to the creation of a limitless source of clean energy, one that would lead the poorest country to progress, that would turn poverty into prosperity. I, along with the rest of the world, was bedazzled by the prospect of nuclear energy, of swords, for once, turned into plowshares.

In 1954, American television sets delivered a live broadcast from Shippingport, Pennsylvania, as President Dwight Eisenhower, in Denver, waved his "radioactive wand"—a neutron source—to set off a signal that triggered groundbreaking for the U.S.'s first reactor for purely commercial electricity generation. Three years later, Shippingport started generating power. The chairman of the Atomic Energy Commission forecast that the Shippingport plant would introduce a new age in which nuclear energy would be "too cheap to meter," a bottomless cornucopia to satisfy even the energy appetite of the American people, who had been doubling their electricity consumption every seven years since the beginning of the century.

The company running the plant, however, soon became aware of the costliness of nuclear-generated electricity. No matter. The government had agreed to pour funds into Shippingport as a lure to the nuclear-energy industry; taxpayers were picking up losses and would continue to do so, according to the plan, from the plant's "cradle to grave."

From the fifties through the mid-seventies, more nuclear plants were ordered in the United States. By the mid-eighties, the United States was generating fully one fourth of the world's atomic electricity. Yet, just as the Shippingport plant had proven from the start, the promise of cheap atomic energy was a false promise.

For decades, physicist and Nobel laureate Henry Kendall, a member of the Cousteau Society's advisory board, has indefatigably scrutinized every detail about nuclear-power generation, risk, and cost. Although fission of uranium, per pound, produces one million times the energy produced by combusting coal and oil, he says, nuclear power is by far the most expensive energy source, due to the astronomical costs of constructing plants strong enough to withstand constant irradiation, not to mention the radioactive onslaught of possible accidents. Kendall has calculated, for example, that a large nuclear power plant produces the amount of long-lived radioactivity that would be released by the detonation of nearly two thousand Hiroshima bombs. Building a plant staunch enough to contain these lethal by-products has cost more than four times the price of building a coal-fired plant. To pay for the new plants, industry saddled consumers with electricity rate hikes of as much as 50 percent.

Then, American energy consumption plummeted. Orders for more than one hundred nuclear plants were canceled.

By 1985, *Forbes* magazine reported that "the failure of the U.S. nuclear-power program" was "the largest managerial disaster in business history, a disaster on a monumental scale." By the 1990s, construction of the last three plants being built in the United States—on which about $8 billion had already been spent—was called to a halt. Cancellations and closings spread worldwide, with a news photo of a nuclear plant in Germany—which had been built at a cost of 232 million deutsche marks and had operated only eighteen days—being demolished at a cost of 289 million DM more. The headline: "How Much a Watt?"

Still other, less-publicized atomic-power costs will also fall to the public—and, worse, to the coming generation and to even more-distant descendants who will have neither consumed the energy in question nor participated in the ill-begotten decisions to generate it.

Industry, for example, has not picked up the costs of full insurance for nuclear-plant accidents. If a plant's cooling system were to malfunction, the unmitigated, intense heat generated by its tremendous radioactivity could melt the reactor's core; molten radioactive materials could rupture through containment structures and erupt into the open, killing tens of thousands of people and contaminating thousands of square miles. The deaths, destruction, and devastation of such an accident would dwarf the global experience with Chernobyl. Estimates of the resulting costs, in the hundreds of billions of dollars, boggle the mind. Is it merely possible or probable that so serious an accident could occur in the United States? In 1985, a year before Chernobyl, a commissioner of the U.S. Nuclear Regulatory Commission testified to Congress that "given the present level of safety . . . [of] the operating nuclear plants in this country, we can expect to see a core meltdown accident within the next twenty years."

Who will pay? In the United States, home-insurance policies do not cover losses incurred because of nuclear accidents. The U.S. government, moreover, has capped industry's liability; if a severe accident causes damage that exceeds the limited dollar amount for which industry is now deemed legally responsible—as experts contend an accident could—either taxpayers or the victims themselves will be left to pay for the rest.

Chernobyl has provided a mild preview of how much consumers, industry, insurance, and government could have to reimburse in case of accident. The Ukraine could not come up with anything close to the billions of dollars needed to close the Chernobyl complex, resettle populations, recompense personal losses, treat farmland soils for radioactive contamination, provide medical care. Nine years after the accident, the director of the Radiobiology Institute of Belarus's Academy of Sciences confirmed that precancerous thyroid conditions were widespread in Belarus. "This is on a mass scale," he said, "several million kids who might develop thyroid cancer." As the people's radiation doses mount through food grown in contaminated soils, so too do cases of congenital malformations and cancers of thighbones, lungs, and soft tissues, particularly among the young. Children in utero at the time of the accident and born after it, sometimes handicapped, often abandoned, are now shuttled between orphanages and foster homes. These children pay the price of nuclear power—one that those who calculate in cash alone cannot begin to understand.

For those who are, however, compelled by the bottom line, the cost that will catch them most unaware must be paid for the job of dismantling all the world's nuclear power plants after they've reached their mandatory retirement age. Everyone in government, military, and industry knew from the beginning that, by virtue of being constantly irradiated, reactor walls would grow brittle, that each entire plant would have to be disassembled, its parts disposed of, after operating only thirty years. From the perspective of officials who launched nuclear power in the fifties, the projected plant lifespan of three decades postponed worries for a suitably long time. Today, both the vantage point and the judgment have changed. To those who must pay for the mistakes made in the past—us—the future is now and the problem is ours.

In the late 1980s, the first U.S. reactor to generate civilian electricity, Shippingport, became the first commercial reactor to be disassembled and buried. This time, alas, Dwight Eisenhower wasn't around with his magic wand. During the five-year dismantlement project, workmen wore heavy shielding to protect against fatal radiation doses as they used picks and shovels to tear the plant apart, scrubbing radioactive walls by hand and fleeing to safety at the frequent shriek of radiation-monitoring alarms. To find such workers—dubbed "glowboys"—the industry has had to rely on recruitment firms that conduct psychological screenings, as one manager put it, "to weed out the abnormal and those who might go off the deep end."

Shippingport's 770-ton reactor vessel, encased in concrete, was finally lifted by crane to a wheeled transport, which in turn drove it to a barge, which in turn chugged 8,100 miles down the Ohio and Mississippi rivers, to the Gulf of Mexico, along the Panama Canal, up the Pacific coast to Washington state, then inland on the Columbia River to the government's Hanford radioactive waste dump in Washington state. There trucks carted the reactor to a trench in a desolate area, where it will remain radioactive for the coming thousand centuries. More than one hundred trucks loaded with Shippingport's radioactive rubble, even the tools and clothing the workers used, also made a cross-country voyage to the Hanford site for special disposal.

Officials involved with the Shippingport dismantlement hailed it as a success. In fact, the challenges that government and industry managed to evade in this introductory dismantlement swing like swords of Damocles above all the dismantlements that await. Shippingport, a relatively small reactor at 72 megawatts, could be transported in one piece. What dangers will be encountered with the worlds' 1,000-megawatt reactors, which can be transported to burial sites only after being carved apart, in a highly dangerous, remote-controlled technological operation? No spills, no train wrecks, no freeway collisions, no barge foundering, no terrorist attacks occurred during Shippingport's dismantlement and long-distance voyage; how will authorities deal, in the future, with radioactive exposures caused by an accident during transport, or even with unsettled radioactive dust at the site? Because, from Shippingport's inception, the government pledged heavy subsidies for its costs, U.S. taxpayers dug into their pockets for its dismantlement. Who will pay for the dismantlement of commercial plants—some of which are ten to twenty times larger? Who will pay, for that matter, for the ultimate dismantlement of the five hundred reactors now online or under construction around the world? The U.S. government buried Shippingport's reactor on government property only because of its responsibility agreement; it has no room, at the Hanford site, for commercial reactors. Who will offer their property for burial of those five hundred reactors that we've built and are rebuilding?

The astronomical cost of dismantling a plant has prompted nuclear industries and governments to propose other, cheaper ways with which to deal with all the disintegrating reactors we've inherited from the past decades: passing them to the future. The first plan, an option they call "entombment," would cost less than half as much as dismantlement. The scheme involves encasing the offensive reactor vessel in reinforced concrete, then guaranteeing that the hardened shroud will endure through the lives and deaths of generations and nations, a bequest to the fortieth century, testament to the improvidence of the twentieth. Another proposed option, "safe storage," entails hiring guards for a hundred-year vigil outside dilapidated plants. No one can promise, of course, that when the guards finally go home anyone on-site will have architectural drawings of the interior in order to start planning a dismantlement. But the proposal has an irresistible financial lure. The companies don't have to worry much about footing the bill for ultimate demolition. After a century they mostly likely will have long since disappeared.

* * *

The question of who will pay to dispose of reactors—especially if the answer is our children—is a poignant and passionate one. Unfortunately, for the moment, it is also moot. No one can yet resolve the problem of who will pay to dispose of high-level radioactive debris because no one has yet resolved the question of what to do with it.

How clearly I remember the hubris of the technologists at the 1959 conference I hosted in Monaco. I can still hear one French technologist's ringing assurance that the world's failure to have yet conceived of a permanent method of storage presented no fearsome threat: "The nation that produced St. Exupéry must have faith in the future!"

His trust in technologists proved to be misplaced. His colleagues at that 1959 conference estimated how much radioactive waste the world would accumulate by the year 2000. As the century now draws to a close, the actual nuclear waste we've produced totals not twice as much as they estimated, not three times as much, but thousands of times as much as the technologists in Monaco foresaw.

Nor has a safe, permanent method of waste storage materialized. Only a few years after their selection, mountain caverns that technologists singled out for eternal storage have been discovered to be riddled with leaks that lead into water systems. Individual containers of waste swell and split as the decaying refuse inside produces radioactive gas. The need to identify other such unanticipated hazards—*before* proceeding with irreversible disposal actions—has prompted the U.S. National Academy of Sciences to declare that studying storage proposals requires "urgent attention." Nonetheless industrialists urge us to speed ahead with uncertain proposals; every year that scientists spend examining the stability of a proposed waste site means another year that nuclear-power companies themselves must pick up the tab for temporary storage of their radioactive refuse. They do indeed urge us to have faith in the future—blind faith.

Blind faith, in turn, has already lived up to its reputation. For the first decades that we stored atomic waste with interim methods, the fact that no one knew about the ultimate effects led many to nurse a sweet delusion that there might not be any. But now our experiments with nature are beginning to register the negative results. For example, nuclear technologists for years pointed to the capacity of the infinite Irish Sea to swallow up and sweep away radioactive poisons. In 1957, an accident still described as the most severe ever at a nuclear facility in the Western world occurred at Britain's Windscale complex at the edge of the Irish Sea. The fallout required such draconian measures as the destruction of all livestock in the area and the dumping of two million liters of contaminated milk.

The government-owned plant that afterward operated on the same site—with a cosmetic name-change from the sullied "Windscale" to "Sellafield"—was repeatedly prosecuted for illegally discharging radioactive material into the Irish Sea. It was nonetheless permitted to add to the sea's cumulative contamination by dumping "legal" discharges of approximately three metric tons of uranium annually.

We forget such practices, but the sea remembers. Today the Irish Sea is known as the most radioactive sea in the world. Salt marshes of the area's estuaries have become reservoirs of radioactive contaminants; government officials warn that construction along the coast could unsettle radioactive silt. Both plutonium and cesium have been detected in sheep that have grazed in the area. Scientists contend that, despite the present world ban on ocean dumping, the radioactivity of the Irish Sea will continue to climb; some of the radioactive compounds now contaminating the water naturally decay into compounds even more biologically hazardous. We cannot rewrite this past; we cannot reverse the damage.

Even the most secure storage methods, being only provisional, have faltered as the years have passed. Energy plants' spent fuel rods, laden with their deadly plutonium and U235, pose the most frightening problem. Even those few plants that have already been dismantled were taken apart and transported only after their fuel rods were removed; the most highly radioactive substances on Earth, fuel rods cannot be buried even in the few appropriate sites available for obsolete reactors. For the time being, the U.S. nuclear-power industry stores its spent fuel rods near its operating reactors, in water tanks that are already almost full. France, like many other nuclear countries, reprocesses spent fuel rods to remove some of the intensely radioactive waste, then mixes what is left with molten glass and forms it into logs to be stored—somewhere, someday, when a site is found.

Because Russia has reorganized the former Soviet secret police agency and given its military a new role, the responsibility for guarding weapons-grade fissionable materials somehow fell between ministries. Thus the vast Russian republic is dotted with labs and energy facilities that stow fissionable materials in buildings sometimes secured only by an easily vaulted fence or lacking even so much as an alarm. Incredulous international inspectors have found, throughout the Ukraine and Kazakhstan, highly enriched uranium casually left in rooms with open access.

If the problem with industry is that its personnel don't know what to do with nuclear waste, the problem with the military is that *we* don't know what they do with nuclear waste. The official stamp of "national security" has served not just as an efficient way of preventing a nation's enemies from learning state secrets but also as a convenient method of preventing its own citizens from learning of military negligence and impending, even ongoing, danger.

Nearly every U.S. plant participating in bomb production has been seriously contaminated. The U.S. Hanford reservation was originally developed in south-central Washington state to produce plutonium for weapons. Officials routinely discharged such great amounts of radioactive waste in the forties and fifties that they exposed unwitting citizens of the country's Pacific Northwest to major radiation doses. Around the world, people were shocked to discover that American military Strangeloves had conducted secret studies in which they irradiated unsuspecting human beings. And when the experimenters wanted to serve their subjects some radioactive meals, where did they find the tainted foodstuffs? They simply went fishing in the Columbia River, so severely had its waters already been laden with radioactive phosphorous from the Hanford nuclear complex. Even today, nearly half of the single-shell storage tanks in Hanford's dump, which once each contained more than five hundred thousand gallons of high-level liquid waste, are leaking. Cleanup of the Hanford site alone is estimated to cost $50 billion. Experts give a chilling prediction of when work will start: "Once the appropriate technology is invented."

What of irresponsible waste-storage practices in other countries? For most of them—even supposedly "open" societies—who knows? It took the fall of the Soviet empire for the Russian public finally to receive information about often slipshod, sometimes absurd, and usually deliberate actions regarding radioactive waste and contamination. In 1994 Russian officials revealed that about half of all the nuclear waste generated from the beginning of the Soviet nuclear program had been injected directly into the earth near the Volga, the Ob, and the Yenisei rivers. Radioactive materials have already seeped "a great distance." As the world's people wait to discover whether the irradiated materials will reach the Caspian Sea and the Arctic Ocean, newspaper commentators gasp that this rape of the earth constitutes the most careless nuclear practice ever—3 *billion* curies of injected radioactive waste—as compared with 50 million curies released during the Chernobyl accident, 500,000 curies discharged in the forties and fifties at the American Hanford reservation, and 50 curies released at Three Mile Island. The greater shame: These are not isolated incidents but cumulative contaminations.

* * *

And so, with no permanent disposal method, and with no guarantees over the safe custody or even accounting of their plutonium and highly enriched uranium stores, what do decision makers intend to do next?

Make more.

Nuclear-energy officials throughout the world have announced plans to quadruple present stores of plutonium for energy plants despite the fact that it could be used for weapons. They plan to achieve this goal in the closing decade of the twentieth century and the dawning decade of the twenty-first.

Whatever for?

The nuclear industry's technologists considered two facts: First, conventional nuclear-energy reactors depend on uranium mines for fuel; and second, by burning that fuel, the plants produce presently indisposable plutonium and $U235$. The nuclear technologists had an idea. Why not separate out the plutonium and uranium from waste with the technology called "reprocessing"? And why not then convert nuclear-energy plants to run on the salvaged materials, processed into a new "mixed oxide" fuel? Technologists also proposed another idea: They could jump-start a veritable "plutonium economy" with their "fast breeder reactor," which not only burns the plutonium-uranium mix but also operates at such high degrees of fission that it actually creates more plutonium than it consumes—thus advertised as spewing out its own infinite fountain of fuel, the Ponce de León discovery of the new nuclear world. (Or so it appears. Be reassured: There will never be an exception to the Second Law of Thermodynamics.)

Physicists, disarmament negotiators, intelligence operatives—all expressed their astonishment that anyone could urge a plutonium economy on a world threatened by plutonium proliferation. Just one country with just one single breeder and one single reprocessor can burden Earth with an additional 8,800 pounds of plutonium every year. A country that sells its reprocessing services, moreover, extends an invitation to terrorists every time a foreign client ships spent fuel for reprocessing and every time it ships separated plutonium back to the client.

A plutonium economy also introduces perils other than proliferation. Because a breeder uses fast neutrons, it operates at extremely high temperatures; the intense heat requires cooling with liquid sodium, posing the relentless threat of fire. Worse, a breeder's makeup is such that an accident could detonate an actual low-grade nuclear explosion, disgorging far more radioactivity than even the worst-case accident conceivable at a conventional reactor site.

As early as 1976, when I was asked to speak to protesters of France's Superphénix breeder, I arrived to discover a virtual tidal wave of people who had come to voice their opposition. In subsequent years, the Superphénix breeder fulfilled some of the protesters' most ominous prophesies. Leak followed leak in its cooling system, an enormous danger given that the liquid sodium coolant is extremely toxic, flammable, and both highly explosive and, once it's been used, highly radioactive.

Next, economists added their voices to the swelling outcry against reprocessing nuclear waste and burning the separated plutonium as fuel. While conventional nuclear reactors are far more expensive to build than power plants burning other forms of fuel, breeders are in turn significantly more expensive to build than even conventional nuclear plants. Superphénix, for example, cost more than two and a half times as much to build as a conventional atomic-energy plant. Reprocessing spent nuclear fuel is also a costly process that in turn produces expensive fuel. Finally, breeders and reprocessors were introduced to provide plutonium as an alternative fuel when prices for uranium were expected to soar. Instead, uranium prices plunged—by 90 percent. Today cheap uranium gluts the market and is expected to remain in oversupply for decades.

While the threat of nuclear disaster had not motivated much action, the threat of economic disaster did. The United States dropped financing for breeders in 1983 and less than ten years later abandoned reprocessing except at a few of its research facilities.

Plans began to founder for a European cooperative project, in which Britain, France, Italy, Germany, and Belgium had joined to finance and construct four commercial fast breeders and produce reprocessed plutonium for fuel. The Italian parliament canceled its participation. Germany's utilities voiced their willingness even to pay penalties to escape their plutonium contracts. Germany also canceled plans, despite decades of investment, to build breeders within its own borders. In 1986, the British House of Commons Select Committee on the Environment prepared a scathing report on its country's foolish heavy investment in reprocessing. Shortly thereafter, Margaret Thatcher privatized electric companies, and the new utilities wanted nothing to do with nuclear energy when they discovered that the British government had greatly understated costs. In 1990, meanwhile, France closed the accident-prone Superphénix itself after a heavy snowfall brought down part of the roof.

As breeder projects were canceled, the reprocessors assigned to fuel them soon began to outpace demand, and a surplus of separated plutonium began to mount at the sites, presenting the reprocessor executives with the expense, as well as the menace, of storing it. When in 1993, after years of having shipped its radioactive waste to France for reprocessing, Japan brought back its first shipload of separated plutonium, an international furor erupted. For fear of terrorism, several nations en route refused to allow the plutonium-laden ship, unaccompanied by military escort, through their waters. At the same time, Japanese financial experts expressed embarrassment that their nuclear fuel cycle was producing the world's most expensive electricity. Soon Japan announced that it would postpone plans for its projected chain of breeders and reprocessors on its home territory. Various experts, including the U.S. White House science adviser, wrote eulogies for the plutonium economy, denouncing plutonium as being "ruinously expensive" and even as having "negative value." Plutonium, wrote one commentator, is "toxic fool's gold." I believed that the breeder reactor was finally dead. I was wrong.

In April 1994, just weeks after Japan promised to postpone its planned breeder and reprocessor projects, the country reversed course and activated its own first fast breeder—dubbed "Monju," after the god of wisdom. Japan proposed to use Monju as a prototype for other breeders that it planned to build around its archipelago. The country has already accumulated copious stores of plutonium to feed all its projected unbuilt breeders; nonetheless, having invested billions in French and British reprocessing plants, Japan plans to import two more shipments of separated plutonium annually until 2010. Japan has moreover broken ground for a first multibillion-dollar reprocessing plant of its own.

Russia too has mapped out plans to reprocess nuclear waste to recover plutonium; the new republic already operates reactors burning the reprocessed plutonium-uranium mix. Moreover, Russians insist that, as they disassemble their arsenals of atomic bombs, they will not seek a safe permanent disposal site for the liberated fissionable materials but will instead hoard it, despite crippling storage costs and the potential for theft and environmental calamities, in order to fuel a new generation of breeders of their own.

France, ignoring scientists' warnings that its star-crossed Superphénix was "ill conceived and dangerous," defiantly reopened the breeder in 1994. Although France already had attained the dubious status of the world's most prolific producer of plutonium reprocessed from nuclear-fuel wastes, it planned to expand its reprocessing potential, envisioning a time when all the country's 900-megawatt conventional reactors will be converted to run on the plutonium-enriched uranium mix and when the country will separate out so much plutonium it can feed them all itself. As for its contracts to separate plutonium from the spent fuel of foreign customers: France has stipulated that it will not perpetually store the leftover, highly fissionable wastes. Having ignored world approbation for its first high-seas shipment of plutonium to Japan, France now plans to ferry boatloads of reprocessing wastes around the globe.

Britain enthusiastically entered into competition with France for these inglorious distinctions, inaugurating, in 1994, the largest reprocessing plant in the world to that date. With no domestic need for more plutonium, Britain overrode seventeen years of opposition from farmers, fishermen, and others in the area of Dounreay, Northern Scotland, who will be exposed to the plant's potential radioactive discharges and who fear the possibility of accidents if flasks of foreigners' spent fuel are unloaded in their busy port and pulled by train through town. How has the United Kingdom proposed to return the plutonium it separates from clients' wastes? By air-expressing it to the nations of origin, via up to two hundred international flights annually, each plane loaded with ten to sixteen tons of plutonium oxide, taking off from one of the United Kingdom's most inhospitable, wind-whipped coasts.

* * *

Surely there must be some reason for this manic plutonium output, one weighty enough to counterbalance these ponderous dangers. In fact, officials eagerly and even passionately defend the benefit for which, they say, they submit their people to plutonium risks: On an overpopulated Earth, individuals crave, and deserve, more energy. Resource-poor countries like France cannot "waste" plutonium, asserts the president of France's reprocessing firm; just one ton of plutonium, he argues, packs the energy of one million tons of petroleum. The head of Russia's nuclear regulatory agency asks plaintively, "Will our children and grandchildren be thankful to us when they experience a lack of fuel in the future?" In the Ukraine, authorities have even continued to operate the ramshackle reactors on the Chernobyl site, despite the fact that, according to a U.S. intelligence report, "conditions at the Chernobyl plant are in many ways worse than those that existed prior to the disastrous 1986 accident"; Ukrainian officials explained that they had to accept the risk of running the damaged plant because their people, so many of whom already lead a scant existence, cannot spare the power produced by the complex. In China, the electric-power minister explains that a full tenth of his own people—one hundred million rural Chinese—have no electricity; even industrial centers like Shanghai suffer blackouts. Officials from various other Asian countries, virtually bereft of oil and natural gas, undergird their arguments for pursuing nuclear power by citing anxieties about maintaining even a basic subsistence living should their imports of fossil fuels ever be cut off.

Yet responding to overpopulation by supplying nuclear power—persuading the poor to buy the most expensive energy on Earth—does not solve world problems; it compounds them. "Energy needs" seems less an explanation for nuclear proliferation than it does an alibi.

If officials feel so urgently about supplying energy, for example, why do they show such indifference toward safer means of meeting those needs? After touring Bulgaria's Koslodui nuclear plants, my team concluded that such common-sense conservation measures as the use of electricity meters and thermostats could so significantly reduce squandered electricity that energy consumption would decrease to match supply. The international experts who evaluated Chernobyl proposed the same remedy for the Ukraine. This simple solution did not stop France and Britain from urging the impoverished Bulgarians instead to purchase new nuclear reactors—with all their concomitant costs of construction, eventual dismantlement, and possible accidents—in order to continue generating enough power to waste.

Moreover, if the nuclear establishment's commitment to supplying power for the disadvantaged is so strong, why does it show such striking disinterest in finding alternative ways of providing clean, limitless energy? Even the five hundred nuclear reactors now operating or being built worldwide could not begin to match the power of the sun, which bathes the planet with so much energy that it would take no less than *one hundred million* nuclear power plants to match it. Already we possess the means to harness some of the energy present in Earth's natural forces— in the sun, the winds, the waves, the heat rising from geological depths, even crops that can be transformed into fuel.

Industry's proclaimed concern for the energy-deprived becomes yet more suspect with a review of its actual customers. Why do nuclear-industry analysts proclaim their preference for clients from oil-rich nations of the Middle East? The satisfaction that comes from helping the energy-deprived poor, apparently, does not compare to the satisfaction that comes from nuclear sales to such resource-rich customers as Iraq, Libya, Syria, Turkey, Egypt, and Iran. Nations whose oil reserves run almost as deep as their international resentments and enmities are not buying nuclear reactors to warm the kitchen hearth. Our nuclear patriarchs are not motivated by altruism, concern for energy supply overall, or concern for the economy as a whole. They make impromptu nuclear exchanges for political expedience and single-sector market gains.

Everywhere one sees the evidence that industry proliferates fissionable material in order to bolster a few teetering companies for the short term by compromising global safety for the long term. France pursues the plutonium economy at least partly because the country has so extensively overinvested in nuclear power that cutting its losses and leaving the nuclear arena could aggravate national unemployment and afflict the steel and construction industries.

In the boom-and-bust cycle of nuclear-reactor construction, languishing nuclear companies have also tried to foist their fiscal woes onto the poor. As one IAEA nuclear-power division spokesman put it: "Export markets are looked upon with increasing interest as a possible way to avoid painful industry restructuring." Vendors have enthusiastically converged on developing countries that have no proliferation safeguards to interfere with sales. Nuclear salesmen have even lured their customers from developing nations into purchases by throwing in such sweeteners as credit arrangements and offers of free atomic technical expertise. Developing nations, happy to take advantage of the deals, in turn have demanded reactors so big, so sophisticated, and so inappropriate to their needs that even the vendors have thought of no credible energy excuse for their sales.

Politicians also see the attraction of the quick nuclear fix, offering fissionable materials and nuclear technology as they would flowers and chocolates in their efforts to woo whatever nations strike their passing fancy—past and potential enemies included. The gold rush for the wide-open East Asia market opened with an industry conference held in Hiroshima itself, with the stark, spindled scaffolding of the city's incinerated Peace Dome serving as the lurid backdrop for the chief of British Nuclear Fuel PLC to announce: "It's springtime in Hiroshima, the start of a new era—the rebirth of an honorable concept" of nuclear energy.

Entries on the list of nuclear goods bartered for immediate profit and political gain are tiresome to read day after day in the headlines, but terrifying to contemplate from the appalling perspective of hindsight.

○ Beginning as far back as the fifties, France supplied then nonnuclear Israel with a reactor that, though poorly designed for power production, was particularly useful for weapons making. Despite the fact that its customer requested a hidden underground workspace, despite Prime Minister David Ben-Gurion's peculiar public announcement that the French were building him a textile factory, France went on to supply Israel with a reprocessing plant with which it could extract plutonium for bombs.

○ The U.S. traded atoms for Afghanistan. When the Soviets invaded Afghanistan in 1979, threatening the security of the Persian Gulf—and of U.S. oil imports—the United States sought Pakistani support. U.S. law expressly forbids sending financial and military aid to nonnuclear-bomb countries that import unsafeguarded uranium-enrichment technology. The quick fix: The U.S. Congress and two sequential U.S. presidents exempted Pakistan, winking at its unsafeguarded reprocessing project and approving billions of dollars in aid, as well as a commitment to sell Pakistan military goods. According to proliferation experts Gerard Smith and Helena Cobban, writing in the journal *Foreign Affairs*, "The Carter and Reagan administrations and their counterparts on Capitol Hill decided that rolling back the Soviet presence in Afghanistan superseded the U.S. interest in preventing nuclear proliferation. They did this even though there was little or no public debate over the relative value of these two goals."

○ After Canada sold a reactor to India—Pakistan's mortal enemy—the United States again sidestepped its laws and sent India, which refused to sign the NPT, nuclear fuel. Today U.S. officials express their surprise that India and Pakistan are adjudged to hover at the brink of nuclear war, with all South Asia described as a "potent tinderbox" subject to devastation by India's estimated thirty to sixty plutonium bombs and Pakistan's suspected covert cache of ten to fifteen.

To the belated astonishment of the nuclear establishment, its customers themselves now hawk atomic products on the "gray" market. Covert-bomb-building Pakistan has provided covert-bomb-building Iran with nuclear technology. Argentina, which had been sold unsafeguarded nuclear material by West Germany, Switzerland, and China, has concluded contracts with Iran to sell nuclear fuel as well as sophisticated technology for handling radioactive material, despite worldwide alerts that Iran is working on a bomb. North Korea has agreed to step up its nuclear cooperation with Iran and has reportedly sent its scientists to Iran to share nuclear expertise. Iran, in exchange, financed North Korea's infamous No-Dong missile, which would enable both countries to deliver nuclear warheads to targets six hundred miles distant.

Such second-generation sales have prompted proliferation experts to mutter bitterly about "rogue states selling to rogue states" and even "an international Mafia"—including Iran, Algeria, and Libya—that cooperates and "aims at getting the Bomb for every member." Yet veteran members of the *Club des Grands* themselves routinely break the sales sanctions they expect others to honor. In just the five years between 1988 and 1992, the United States approved 1,500 exports of sensitive nuclear-related equipment, worth $350 million, to eight countries suspected of nuclear proliferation. Given that this period covered the years just prior to and after the 1990 Gulf War, the presence of Iraq on the U.S. list of customers is dumbfounding.

Equally inexplicable are revelations about Iraq's other vendors. In 1975, France sold Iraq, on whom it depended for oil, two reactors and enough enriched uranium for about a half dozen bombs. That the Élysée Palace itself knew the deal was unsavory is apparent in the fact that it kept the arrangement secret from its own people for a year. When the deal was finally revealed to the public, politicians described it untruthfully, representing it as a commercial, rather than a government, transaction and thereby evading debate in the French parliament.

Likewise in Britain, three senior ministers arranged to export lathes with which, they knew, Iraq could produce components for nuclear bombs; the ministers kept the deal secret because they anticipated public antipathy for sales at a time when Saddam Hussein was committing atrocities against the Kurds. The British ministers remained mute even when executives of the manufacturing company that had been secretly permitted to export the lathes were arrested and tried by Britain's uninformed customs and legal system. Germany also supplied Iraq with uranium and machine tools useful in the technology needed to enrich uranium to weapons-grade. Germans continued to sell Iraq nuclear goods after Saddam invaded Kuwait; evidence indicates that Germany even shipped nuclear technology to Iraq, via Pakistan, three weeks after the United States had entered the 1990 Gulf War.

As John Deutch, then deputy secretary of defense and later director of central intelligence, reported in the prestigious journal *Foreign Affairs*, Iraq was found to have been operating a massive covert bomb program—mining uranium ore, enriching it, designing weapons, and involving more than ten thousand qualified technical personnel. Hussein-controlled newspapers in Iraq even asked if the United States understood "the meaning of every Iraqi becoming a missile that can cross to countries and cities." Yet none of these clear expressions of atomic intent stopped one country, thought to be Pakistan, from selling Iraq more uranium-enrichment equipment. Nor did they dissuade a German-registered ship from carrying a shipment of Chinese-produced ingredients for ballistic missiles toward Iraq. Nor did they stop France's state-owned oil companies Elf Aquitaine and Total from rushing to become the first Western firms to contact Baghdad after the 1990 Gulf War, asking for business in return for which they offered their lobbying efforts to lift U.N. sanctions.

* * *

After decades of quick-fix nuclear sales, industrialists and politicians have spawned pandemonium.

Today the IAEA guesses that civilian plutonium already separated from nuclear-energy plant fuel and usable for bombs could total more than one hundred tons and that this total is stocked by some twenty-two countries around the globe. No international law requires fundamental health standards for the storage or handling of this plutonium. Current methods of transporting these materials have been judged to endanger human populations; those who ship plutonium by sea are not even required to seek military escorts. Accounting methods—efforts just to keep track of how much plutonium is stored—are variously cavalier, inept, or altogether nonexistent. In Britain, experts calculated that more than two tons of possibly weapons-grade plutonium in the country's Sellafield stockpile had "gone missing" without explanation. Authorities at the country's Sizewell nuclear facility contended they had erased computer records and thus did not even know how much plutonium their facility had produced. Japanese nuclear officials admitted they couldn't find 154 pounds of plutonium and submitted that the missing amount—enough for nine bombs—was probably lost inside their fuel-making plant in the form of dust. U.S. experts estimated that seven bombs' worth of plutonium was missing at the Rocky Flats nuclear facility. Officials suggested it was probably "lost in the pipes." The U.S. Department of Energy has admitted that it cannot actually account for America's full inventory of plutonium. In fact, it has been revealed, the U.S. government has even misplaced warheads.

Proliferation experts look back at the days when terrorists posed no nuclear threat because bomb making was a successfully guarded secret. When the secret was out, would-be bomb builders were foiled by the lack of plutonium. Today, according to experts from Clark University, both highly enriched uranium and plutonium "may be easily stolen in significant quantities from the mature fuel cycle, national and international safeguards notwithstanding . . . The theft of weapons-grade material from a mature nuclear-power program is a catastrophic event . . . a would-be nuclear terrorist need only steal the right shipping container and learn the detonation, where formerly it was thought he would have to embark on at least a miniature Manhattan Project."

The nuclear apologists who dismiss the possibility of nuclear thefts in the future are well advised to review the past. The CIA first reported that it believed Israel had already produced nuclear weapons, "based on Israeli acquisition of large quantities of uranium, partly by clandestine means"—an announcement that brought to mind the two hundred pounds of highly enriched uranium that mysteriously went missing from a Pennsylvania plant decades ago. Thieves have also served as the main suppliers of the alleged Pakistani bomb program, reportedly launched by a Pakistani physicist who stole blueprints for an enrichment plant from an engineering company that had employed him in Holland. A Dutch court sentenced him—in absentia—to four years in prison; but the convicted thief was already back living in Pakistan, serving as chief of its nuclear program.

Both Russia and the United States have allegedly cooperated in keeping discussion about more recent thefts out of the public forum. In 1993, security police confiscated almost eight hundred pounds of reportedly weapons-grade contraband uranium in the Black Sea port of Odessa, Ukraine. A janitor is said to have strolled out of a Russian nuclear plant with three kilograms of highly enriched uranium stuffed into an industrial mitten; the janitor was discovered missing before the uranium was. During U.S. congressional hearings in the summer of 1995, a witness testified to a single theft that involved "at least one senior Russian government official; a senior official at a nuclear institute . . . an organization believed to be linked to the KGB or what used to be the KGB; organized crime mobs in Russia and Lithuania; and very likely an arms merchant with a history of dealing with Middle East states and terrorist organizations." Senator Richard Lugar, who chaired the hearings, observed that the risk of a nuclear bomb exploding within U.S. borders had increased rather than decreased since the end of the Cold War and that government "has not even begun to appreciate U.S. stakes in the matter."

And still insouciance and recklessness rule the day. The maximum sentence for nuclear smuggling is no more than five years. Interpol has historically considered plutonium as nothing more than a "health threat."

No international agency regulates or even monitors nuclear trade.

Perhaps it is time to admit that no one ever will. Perhaps it is time to reject the explanation that officials impose the risks of proliferation on our world because we can't give up nuclear energy. They impose the risks of nuclear energy on our world because they can't give up the bomb.

* * *

Is the bomb worth it? Has it in fact guaranteed the peace and security that nuclear potentates insisted it would?

- ○ The atomic bomb has failed to guarantee peace. The notorious "deterrent" power of the atomic bomb stopped neither the Chinese

Communists from invading mainland China nor North Korea from invading South Korea. The menace of two nuclear powers engaging in ground fighting against each other deterred neither China nor the Soviet Union in 1969. Britain's bomb did not terrify Argentina into staying off the Falkland Islands, nor did it prevent the Irish Republican Army's bloody assaults on British citizens in their own homeland. The American bomb did not dissuade the Soviet Union from backing North Vietnam any more than the Soviet bomb stopped the United States from backing South Vietnam.

The terror of nuclear conflagration attached to Israel's possession of the bomb did not prevent Saddam from lobbing Scud missiles at Tel Aviv. It only prevented Israel from responding.

○ The atomic bomb has not guaranteed us security. The mere existence of sophisticated atomic arsenals, in fact, increases the possibility of accidental nuclear warfare. Today we are threatened by a "significant probability" that nuclear war can be set off accidentally, especially in times of crisis, according to participants of a conference sponsored by the U.S. Brookings Institution and the Center for Science and International Affairs at Harvard University. These experts have warned that the systems controlling, targeting, and commanding weapons have grown so complex that they are increasingly susceptible to false alarms, computer failure, and human error. U.S. military forces received one hundred false missile alerts in just the four-year period between 1981 and 1985. Among the sobering real-life incidents: an occasion in which a computer chip failure at U.S. Strategic Air Command signaled an incoming Soviet missile. Before the alert could be canceled, attack warnings sounded at U.S. missile silos, B-52 bomber crews were scrambled, and nuclear-submarine commanders were readied.

○ Nations who savor the prestige of acquiring their own arsenals do not guarantee their homelands peace but rather invite attack, as Saddam Hussein discovered. Retired general Colin Powell, chairman of the joint chiefs of staff in the United States during the 1990–91 Gulf War, described international realities succinctly when he was asked what

North Korea should be told about nuclear weapons. "If we ever think that you're going to use one," the general replied, "or if you do use one, you'll become a charcoal briquette."

o   Possessing atomic energy plants in itself amplifies a nation's peril. Should a single one-gigawatt reactor near Stuttgart, for example, be hit with a single one-megaton bomb, the resulting explosion could devastate "a substantial part of Europe," according to Steven A. Fetter, at the time dean of the School of Public Policy at the University of Maryland, and Kosta Tsipis, associate director of the Massachusetts Institute of Technology's Program in Science and Technology for International Security. "Vaporizing cores of nuclear reactors with nuclear weapons is clearly an efficient way to desolate large parts of a nation."

o   Most significantly, the atomic bomb cannot be believed to guarantee peace when it has introduced the prospect of atomic terrorism. In the United States, an extortionist has already threatened to use an atom bomb to annihilate Boston unless he was paid $200,000. The U.S. government has reviewed more than one hundred threats and mobilized for action regarding thirty of them. All hoaxes. But officials have taken them seriously enough to invest $70 million annually into its special SWAT team of more than one thousand specialists, zealously readying themselves for action on the day the threat is real. So too has the U.S. Department of Defense begun preparing for the fanatic, contracting a study ominously titled *Terrorism 2000.* Coordinator of the report Marvin Cetron concluded that the possibility that stolen fissionable materials will be used is "very high." "An improvised nuclear, biological, or chemical attack on the United States is increasingly probable; perhaps within the next five years," he remarked in 1995. "If the 1993 bomb that went off under the World Trade Center in New York had been wrapped in radioactive Cobalt 60, you wouldn't be able to get within a mile of the place for the next twenty-five years." The new superterrorists, he added, "are motivated by fierce ethnic and religious hatreds. Their goal will not be political control but the utter destruction of their chosen enemies. They are on the scene, right now."

Back in 1976, at the public protest over France's Superphénix breeder, I was asked to speak about my objections to the plutonium economy. I described the atomic future I feared, a future in which proliferation would force nations to form an international police state, the only way possible to ensure public safety. No one, I said, objects to inspections at airports; we willingly open our bags in official searches for airline terrorists. One day people will just as gladly sacrifice their daily privacy in the search for nuclear terrorists. Mail will be opened, telephones tapped, and people will welcome the restrictions because they will welcome the protection. Given the choice—world holocaust or world police state—the answer will be life, no matter how dismal.

*   *   *

There remains only the last and most specious rationale that nuclear potentates offer for behaving as they do. Nuclear history is replete with statements from military and industry authorities that they are acting on behalf of the public welfare. They must reprocess their fuel, operate nuclear power plants in dubious states, stockpile weapons, test bombs—all for our own good: for our security, our energy needs.

Yet the record shows that the good of civilians has taken second place to military and industrial considerations. After the U.S. atmospheric bomb tests at its Nevada site near Las Vegas, American officials told parents that "sunshine units"—their euphemism for radiation levels—were "well within safe limits," even as radiation killed farm animals in Utah and plutonium was found in the soil as far away as Salt Lake City, Denver, and Houston.

I myself have felt not only anguish but indignation ever since a personal friend confided his own firsthand experience with authorities' attitudes. In the summer of 1963, I attended a meeting of the advisory council of Monaco's Oceanographic Museum, of which Dr. L.A. Zenkevitch was also a member. He was a remarkable man. In the Soviet

Union, he served as president of the Academy of Sciences, a prestigious agency that wielded tremendous influence with the Politburo. He was a renowned marine scientist; he spoke fluently about multiple scientific fields as well as about art, philosophy, and literature, and he was a personal friend of my mentor, Professor Louis Fage. He rarely had an opportunity to leave his country, and so after our meeting in Monaco, I invited him to come as my guest to UNESCO's General Assembly, which was convening in Paris.

Zenkevitch and I sat together at the UNESCO conference, listening to endless skirmishes between the two contending teams of superpowers. I had not invited him to Paris, however, merely to be a spectator at such futile games. I really wanted to have the opportunity to speak further with him. It was a specific, awkward subject I wanted to raise with the great man. At the Oceanographic Museum in Monaco, I was one of the directors of the Marine Radioactivity Laboratory, which monitored daily radioactivity in the atmosphere and the rain and was part of the IAEA's network of radiation control centers. When the Soviets had conducted their infamous series of atmospheric H-bomb tests—and when the United States had responded with bomb tests of its own—our counters had flashed continually. The huge graph we had hung on the wall and updated daily had climbed toward the ceiling. I was so distressed by our findings that I wanted to ask Zenkevitch if Soviets had considered the consequent global irradiation before the tests had begun.

I invited Zenkevitch to dinner. I asked the deputy director of the Oceanographic Museum, Commandant Jean Alinat, to come along, as well as Henri Lacombe, of Paris's Museum of Natural History.

Through hors d'oeuvres and the main course, we discussed education, underwater photography, Indian Ocean Year, exploration submersibles. I described the seismic reflection surveys of the Mediterranean seabed that we were completing aboard *Calypso*. I asked every question but the one I now felt to be too confrontational to put before a good and honorable man.

When our waiter brought the ice cream we had ordered for dessert, I knew I had run out of time for small talk. "Doctor," I began uncomfortably, "in Monaco we are measuring atmospheric radioactivity. During the

Soviet tests, the level became extremely high and probably dangerous all over the northern hemisphere. Why? Why all those bomb tests? Did your government consider the repercussions of their program?"

Zenkevitch stopped eating and cast his eyes downward. My heart was drumming. I was thinking that I should never have raised the question. I loved Zenkevitch and feared I had offended him.

Finally, he drew himself up. He spoke softly: "Cousteau, we at the Academy of Sciences were consulted by Soviet officials prior to the tests." He paused. "After careful evaluation, we warned officials that the proposed program of atomic bomb tests could cause the deaths of perhaps fifty thousand Soviet children."

Fifty thousand children. My mouth went dry. Zenkevitch, president of the powerful Soviet Academy, had tears in his eyes. He continued haltingly. "We were told that if the Soviet Union did not test the bombs, it could cost the country many more lives."

I do not remember how long we remained silent as we ate our ice cream or how we parted or what I did the rest of that evening. My mind reeled. If the Soviets really believed that their tests would result in fifty thousand child fatalities within the Soviet Union, how many children did they think would die around the world? The radioactive clouds had set our monitors in Monaco swinging only after they had been pushed by westerly winds almost around the planet. All along the way, toxic dusts had slowly wafted down, or were washed by rain onto the fields of America, Europe, Asia. Cows had concentrated the poisons in milk— the milk that had been drunk by our own children.

It suddenly seemed as if the graph we were tracing in Monaco represented a government-approved daily nursery toll. Those who had been indifferent to death and had issued the order to test would remain unsuspected as criminals; they would probably die and be buried with honors, with no one knowing that they assigned more importance to their military's bombs than to their country's next generation; with no one aware that, in the name of defending their country, they had chosen to risk sacrificing its children.

Some epidemiologists have since said it would have been extremely difficult, if not impossible, for the Soviets at that time to make a fatality forecast. This comment misses the significance of Zenkevitch's avowal entirely. The point is that the forecast *was* made, by the most distinguished body within his country, and that his government, accepting the warning as well founded, had nonetheless decided to proceed.

Should we assume that a story like Zenkevitch's, in which "the good of the people" is not just abandoned but even compromised, was only an aberration of an evil Soviet empire? That such abuses of the past have been eradicated from the present? Consider Chernobyl. Britain's agriculture minister encouraged parents to give their children milk, contending that if an infant drank a full liter at the peak level of radioactivity daily for thirty days, the child would receive no more than just one single X-ray's worth of radiation. The agriculture minister made this pronouncement despite the news, published that very day in his own country by the distinguished British science journal *Nature*, that the average twelve-month-old infant given milk in endangered areas was in fact receiving a radiation dose equal to forty-five chest X-rays. Only after the fact, when children had dutifully swallowed their milk, was the public told that some 250,000 British children had received dangerous radiation doses—unless their parents had taken the precaution of withholding fresh milk after the cloud arrived. Children in highly irradiated areas of Britain subsequently were condemned to a 40 percent greater chance of developing thyroid cancer.

In the United States, critical information has been withheld not only from the electorate but also from their leaders. Five years after the fact, Americans discovered that during the hours of the U.S. Three Mile Island accident, government regulators and industry executives secretly discussed the danger posed by a hydrogen bubble that had formed inside the reactor vessel; if it had ignited, the resulting explosion would have caused an instant meltdown. Yet at the time, officials withheld information about the hydrogen-bubble menace not only from nearby populations but also from Secret Service agents as well as the U.S. president, Jimmy Carter, who walked through the endangered plant and then innocently promised his electorate that he would be "personally responsible for thoroughly informing the American people about this particular incident and the status of nuclear safety in the future."

The facts about radioactive contamination can be hard to come by, but even when citizens have somehow acquired information, they can be subject to those who try to manipulate them into silence. When I protested France's plans to dump radioactive waste into the Mediterranean, the resources of the Oceanographic Museum were investigated and my back taxes audited; everything, of course, was in order.

From the opening curtain of the nuclear drama, however, those whose welfare has been most compromised by the nuclear aristocracy have been the children of the future. What price will they pay for today's decisions to plunge deeper into nuclear energy? How can any mere mortal claim power over the unknowns of the 250,000 years for which nuclear waste may remain dangerous? Can we really believe that all the unponderables and unpredictables of the planet, that tidal waves, geological upheavals, even asteroids will all respect the integrity of our atomic burial sites and power plants? Some 100,000 earthquakes rock the ground each year. Just between 1970 and 1995, the earth's major earthquakes alone took more than 400,000 lives. In 1993, U.S. engineers were thunderstruck to see 120 of their "earthquake resistant" buildings crack in California's quake. In Japan, where building codes are the strictest in the world, an official of Tokyo's Construction Ministry boasted in 1994 that his country would be immune to such destruction in a quake. "Our standards are higher," he said, just one year before more than 50,000 buildings in Kobe crumbled as the earth's fury proved him wrong.

The nuclear establishment has set a sorrier record still with its predictions. When the Soviets supplied extensive nuclear help to the Chinese, they never anticipated that their agreements would be torn to tatters by the time the Chinese tested their first bomb. When Americans supplied nuclear equipment to the shah of Iran, they never anticipated the revolution that would replace him with an ayatollah who considered the United States to be Satan's equal. When Israelis provided maintenance for that Iranian equipment, they never anticipated that within a decade the Iranians would be using it to finish a bomb for themselves and financing missile designs to shoot those bombs at Jerusalem. The inadvisability of counting on stability is possibly best illustrated by the case of Brazil, which served as a principal during the Latin American Nuclear Free Zone Treaty talks at the Tlatelolco palace. The ultimate adherence to the final treaty would be thrown into question by the ouster of Brazil's president in a military coup during the course of the negotiations.

Nor can nuclear visionaries be clear-sighted regarding their wastes. The Soviets never anticipated, when they dumped radioactive waste from plutonium reprocessing into the Lake Karachay reservoir, that its waters would evaporate in a drought. So too have Americans been surprised to see the radioactive remains in the thousand cavities they dug for underground test blasts, with no barriers to inhibit movement through groundwater, migrate at "alarmingly fast rates," endangering the water supply of populated areas.

And how has this disregard for public safety affected the public who sees its consequences play out? The psychological impact of the Chernobyl disaster speaks for the atom's bleak contribution to human welfare. The so-called Chernobyl syndrome was officially defined as a hopelessness that eroded the lives of Chernobyl victims who no longer felt trust in their authorities, who feared the absence of information, who feared for their lives and their children's lives. Even a decade after the accident, victims continued to consider themselves a "community of the damned," members of a "victim culture." Their responses eerily mirrored those of some Hiroshima survivors, who continued their lives as living dead; after escaping death from instant incineration and death from radiation disease, they succumbed to a living death, their bodies healed but their hope ravaged, defeated by what came to be known as "leukemia of the soul."

Back in 1946, Louis Mumford took an unflinching look at the nuclear aristocracy. "The chief madmen," he wrote, "claim the titles of Admiral, Senator, Scientists, Administrator, even President. Without a public mandate of any kind, the madmen have taken it upon themselves . . . to corrupt the face of the earth . . . The madmen have a comet by the tail, but they think they prove their sanity by treating it as if it were a child's toy. They play with it; they experiment with it. Why do we let the madmen go on with their game without raising our voices? Why do we keep our glassy eyes fixed in the face of this danger?

"There is a reason," Mumford wrote. "We are madmen too."

*　*　*

We need not remain mad. When Einstein once remarked that he wished he'd never "opened this Pandora's box," he chose his metaphor well. In the legend, written as an allegory for the futility of war, Pandora lifted the lid of her jar and liberated all the evils of the earth. Only hope stayed within.

There hope remains—ready for us to extricate it. Already pockets of people are beginning to shake themselves from their mad anesthetic slumber. When a nuclear power station was to be expanded in the French village of Chooz, women locked up the town's mayor. In Israel nuclear technician Mordechai Vanunu felt such profound antipathy for the bomb program on which he had worked that he provided the London *Sunday Times* with photographs and technical details of Israel's secret nuclear weapons plant south of Jerusalem in Dimona; he is now paying for his passions with an eighteen-year prison sentence and has been in solitary confinement since his capture. Mexicans angered over the construction of the Laguna Verde plant organized seventy antinuclear groups in one year. France's decision to explode test bombs drove two thousand protesters into the streets of Paris, hundreds of demonstrators to the streets of Stockholm, and a thousand New Zealanders to the French Embassy, where they hurled eggs and set fire to a mountain of croissants capped by a French beret. When the people of the American state of Maine discovered their area was short-listed for a nuclear-waste repository, an activist swore, "We have people in Maine who are willing to die to keep the Department of Energy out of here." These, our leaders are learning, are the kamikazes of peace.

To protest the bomb with fire and blood is to pit insanity against insanity. Yet there are other means to achieve noble ends. The populations of nonnuclear nations constitute 80 percent of the earth's population. Their voices are now joined in a chorus from many among those who live in nations that possess the bomb. The choices made by a few politicians in a few nations affect us all. We can stop them from bartering away our lives and our children's lives. We must simply realize that we need no longer surrender ourselves to the mad irrationalities of a few dozen men when we are a few billion people.

Early on, the pacifist and philosopher Bertrand Russell, who joined with Albert Einstein to draft their famous manifesto, warned that "facts which ought to guide the decision of statesmen do not acquire their due importance if they remain buried in scientific journals. They acquire their due importance only when they become known to so many voters that they affect the course of elections." These facts, indeed, have been waiting in scientific journals for us to understand since the opening of the atomic age. Physicists who know and fear the power of the atom have long called for the minimum requirements to rein in the chaos: We must create an international organization to oversee plutonium and highly enriched uranium, and we must endow it with authority equal to the threat. We must demand that all nations release data on their inventories of the deadly materials. We must create and enforce methods of accounting for world inventories. And we must ban the further production of highly fissionable materials. Present stocks hold more than enough for world military and energy requirements.

All these demands, while obvious and essential, are inadequate to solve the nuclear problem. There is only one way to rid ourselves of the threats of atomic bombs and atomic terrorism, only one way of assuring that the costly and fearsome probabilities we recognize today will not leave our children with bankrupting and terrifying realities tomorrow. We cannot merely ban the production of fissionable materials; we must ban their existence, ban plutonium and highly enriched uranium on Earth. A world without atomic bombs, atomic terrorism, and atomic contamination can be achieved only by a world without atomic energy.

Si vis pacem, para bellum, declared the Romans; if you want peace, prepare for war. Throughout history, humans have obeyed the dictum and have failed. If we want peace in the twenty-first century, we should prepare for peace. The leader who uses the bomb will destroy the world. But the one who finds a way to ban the bomb will rule it.

The task before us today lies in refusing to be defeated by what we regard as the impossible and in beginning our struggle for what we know to be essential. The necessary has always proved to be possible. Regarding survival as imperative is far more logical than accepting disaster as inevitable. There will be those who insist that a fissionable-materials ban is unworkable; the response to them remains the same today as it was when Bertrand Russell and Albert Einstein drafted their famous manifesto. "We appeal, as human beings, to human beings," they pleaded. "Remember your humanity and forget the rest."

The Alps rise; volcanoes spew; earthquakes tremble; and as time progresses, so must we. Astronomers say that, one day, a meteor could very possibly collide with Earth. No one has dared propose, of course, pointing weapons toward planet Earth's incoming annihilators; the idea seems so foolish. Instead we point our weapons at ourselves.

We could, of course, ban the production and possession of all fissile materials. All—except for enough to create four strategic weapons. The strongest and noblest among us could guard those for the meteor storm, for our literal Star Wars, for the time when we use cosmic forces where they belong: in the cosmos.

CHAPTER TEN

LIFE IN A BILLION YEARS

DURING my long night watches, I often feel an eerie sense of unreality pervade the solitude, the silence of the bridge. The horizon has long since vanished with the light; the ink-black sea merges with the sky. *Calypso* seems suspended in darkness. From behind, waves surge into fearsome mountains of water, their white crests leering down from overhead. Worries that linger from daylight hours loom into phantoms in the dark. Headlines of the evening news come back to haunt me in the dead of night: Ice caps melt; ozone thins; impoverished masses simmer in bitterness as the wealthy few indulge in luxury and waste. I have learned to shake away night demons as I navigate the sea, but I am less confident about navigating through life, in a world that seems to be pitching and rolling toward social chaos or nuclear terrorism, perhaps only twenty-five years, fifty years—just a generation or two—ahead. The wind moans. So do I.

I find relief even as I find my bearings, by the stars—the same stars that guided me fifty years ago on the *Jeanne d'Arc*, the very stars that enabled sailors of ancient Greece to fix their longitude at a time when Socrates envisioned the human future as a distant golden age. I look toward the east for the break of day. As the stars recede in the rays of dawn, my mood brightens, nightmares fade to daydreams, and I find myself imagining my own golden age, not necessarily a future in which humans have avoided catastrophe, but one in which they survive in spite of catastrophe, a future not just one generation hence, or two, but a thousand years ahead, a million years, even—why not?—in the coming billion years.

For those who measure the future only in tomorrows, talking about the next billion years may seem eccentric. But think: Geologists and paleontologists tell us that the planet has already existed five times that long. Astronomers say that the earth can remain populated about five billion years more if it dies a natural death, when at last our star, the sun, finally expands and incinerates the planets in its system. A billion years, then, represents only a fifth of the possible future. From a cosmic perspective, it is we dreamers who are realists.

Such a span of time seems unimaginable to anyone pounding the pavement of a city. But venture into one of the few vast jungles, the rain forests, any of the dwindling realms of nature that remain unviolated. Perhaps I'm so tempted to imagine a vision of the distant future because I've seen evolution's distant past each time that, while diving, I've gazed across the primeval expanses of the ocean wilderness.

The coral reefs we've filmed in the Red Sea, the Indian Ocean, and the Pacific have played out for us the slow, majestic beat of time. Through the portholes of a deep exploration submarine, the true picture of what a reef is and has been emerges, logical but frightening. The corals, all elaborate lacework on high, change the farther we plunge below; in the deep they turn to huge, lifeless mountains, constantly cascading calcium, weeping throughout all time their tears of sand. The colossal scale of the monuments and the fragility of the vibrant layer near the surface testify to the ways in which time itself has served as a construction material. Industrious little polyps and obstinate, indefatigable calcareous algae used the millennia like brick and mortar, extracting calcium carbonate from the sea, then layering it with years to erect these massive ridges. Buried thousands of feet within lie the fossils: stromatoliths, the first of the reef-building creatures, interred about two billion years ago, when the planet was about half again as young, long before even a fish existed. The colorful sprays of living coral near the surface, like memorial wreaths, mark the barren primordial tombs below.

How can we, who measure our lives' events in weeks and months, conceive of the time span over which those minuscule creatures built their massive coral mountains? Imagine standing close to the Washington Monument. Suppose each yard-long step equals fifty years, almost a lifetime. Forty steps back in time and we reach the birth of Jesus. After two hundred steps, we stand only about two blocks from the monument, but already we find ourselves in prehistoric times. Fifteen miles from the monument appear the first anthropoids. But steps, blocks, even miles measure little of the history of life on Earth; to reach the era of the earliest reef-building creatures, we would have to walk around the entire globe.

The saga of evolution not only defines our place in time but also defines our place in life's fraternity. We living beings were described by Loren Eiseley as cosmic orphans, with no memory of the exploding stars, galactic collisions, primal storms, and erupting volcanoes that conceived us. Yet we cosmic orphans nonetheless are not alone. Our entire family of life, plant and animal, warm-blooded and cold-, all of us are siblings joined by our common ancestors, the first single cells nurtured by earth and sea. The events that took place in the billions of years past can surely guide us in our dreams of the billion years ahead.

We still lack some pieces of the puzzle of evolution; yet we have found enough clues in scattered fossil fragments to help us envision its slow, inexorable path. In fact, the longest geological era—the Precambrian, which spanned about the first four billion years—produced the fewest forms of life. At first, the scorching new planet could not support living beings. Its sky flashed with lightning and solar flares; volcanoes spewed various gases into the oxygen-void atmosphere along with steaming clouds of water vapor. About a billion years elapsed before the first plant cell appeared in this broth—we still don't know how. More than three billion more years passed before life finally made its metamorphosis from single cell to complex animal, possibly when plant cells had taken in enough CO_2 to liberate sufficient oxygen. Combusting the power provided by oxygen, life exploded with variety.

Death periodically swept the planet; not all forms of life could endure nature's abrupt environmental changes or its calamitous caprices. Ice ages, meteorite rains, the warming of the climate, the fall of a giant asteroid—which of these caused the widespread extinctions we see grimly parade across the fossil records? Perhaps a change in the magnetic field temporarily disrupted the Van Allen belt, subjecting all life on Earth to heavy radiation, eliminating many creatures as well as causing an increase in mutations.

Like a real tree, the tree of life lost entire branches at times but continued to flourish and ramify. Evolution tested nearly every innovation imaginable: the mobile and the stabile; soft and hard; skeletal and shelled; internal incubation of eggs and external spawning; weightlessness in the sea and gravity on the earth. The number of nonviable beings born of random mutations was probably several thousand times greater than the number that left their remnants in rock to reach, to teach, our times.

The most vivid trace of evolution, however, is etched not in rock but in living tissue. Somewhere along the line of chance combinations came the awesome accident we call the brain. If the human species survives another billion years, the human brain will probably be as different from ours as ours is remote from the simple organs that allow single-celled organisms to react to stimulation.

In spite of the fact that various animals perceive events in the outside world in various ways, the progressive development of the nervous system along the course of life displays a remarkable continuity. The development of the brain and the development of intelligence both progress on parallel tracks as evolution unrolls up to the primates. There the trend abruptly breaks into the schism between chimp and human. I have seen some toothed whales behaving more intelligently than apes. Our closest evolutionary counterparts may not be primates, but animals at the very tip of another branch of the family tree: marine mammals.

Yet our intelligence comes from more than our singular human brain; intelligence also arises from the laboriously accumulated and meticulously recorded sum of collective experience, the history, the heritage, of what we call civilization. Records of feral children, adopted in the wilderness by animals, testify to the value of the civilization of which they were deprived. When found and gathered back into the human community, these children have walked on all fours and have failed to absorb education to average human levels. Having become wild animals themselves, they probably have developed the intelligence of wolves. Even with his complex brain, the human being does not realize his human potential if separated from the human species.

Civilization requires intelligence and more. The dolphin and the orca have intelligence, as well as voices, social communities, and at least thirty years of life expectancy to use and develop their gifts. Yet even in Leo Szilard's fantastic short story "The Voice of the Dolphins," his dolphins win their Nobel Prizes only after their subordinates, the human beings, have literally lent a hand. These sleek marine mammals, smart as they are, lack manipulators with which they could use tools.

An octopus has more than enough manipulators and probably enough intelligence but has no voice and lives only two or three years.

The human being can modulate his voice to communicate with others; he has hands that can use tools, build cities, record and bequeath knowledge from his generation to the next. He enjoys a life span long enough to put his gifts to good use. To build a civilization, one human being needs only another.

Philosophers and poets alike have through written history celebrated the advent of the mind. "Man is a reed, the most feeble thing in nature; but he is a thinking reed," wrote Pascal. "The entire universe need not arm itself to crush him; a vapor, a drop of water, suffice to kill him. But when the universe has crushed him man will still be nobler than that which kills him, because he knows that he is dying, and of its victory the universe knows nothing." With those words Pascal crowned the human being as evolution's royal heir, elevated above all other creatures.

Yet to what end does the human being apply his extraordinary evolutionary windfall, his ability to think his way to harmony and peace? Our abuse not just of other human beings but of the natural world as well—our use of intelligence to engineer nuclear warheads and other engines of destruction—makes one wonder if the complex human brain is itself a lethal mutation.

The possibility that our cerebral blessing will bring about our final damnation arose only in recent human history. For most of his existence, *Homo sapiens* had to struggle against nature to survive. A creature with almost no defensive or offensive weapons, the human had access to little more power than he could cajole from a mule. Suddenly, stumbling upon the secret of fossil fuel, he found himself the unexpected ruler of the planet. He has not yet mastered his supremacy. He does not understand that his survival now depends not on the conquest of nature, but on the protection of nature. Man has ascended to his level of incompetence.

The human being uses modern power with a primitive attitude. The "nature, red in tooth and claw" side of his animal heritage always lingers near his surface; with the additional, limitless power of brain, language, and hands, the human being has liberated himself from natural law but has propelled himself into planetary anarchy. Drunken with hubris, humans have begun to strip the planet of as many species as natural disasters have. As we on *Calypso* voyaged in the Antarctic, we saw a simple but somber illustration of this truth. Diving hundreds of times with Aqualung and diving saucer, we watched how vegetal plankton—which evolved 2 billion years ago—were fed upon by tiny crustaceans, which proliferated 600 million years ago, which in turn were eaten by dolphins and whales, which perhaps appeared 50 million years ago. Splayed across the bottom: a profusion of immense whale bones in endless rows, skeleton after skeleton, mortal remains of members of a magnificent species that has roamed the seas for millennia, reduced to a fragile few by the human newcomer on the planet. We Queequegs not only eradicate the whales but eradicate one another as well, spending far more time applying our incomparable ingenuity to war-making than to peacekeeping.

Albert Einstein described the paradox created by the progressive development of human intelligence: "The level of thinking we have done thus far creates problems that we cannot solve at the same level as the level we created them at." Nonetheless, as Einstein himself demonstrated, the singular beauty of the human brain lies in its potential to accomplish tomorrow even that which it does not imagine today. Given our remarkable gift of thought, we have always somehow figured out a way to save ourselves from ourselves; our brain is the computer that programs itself. Perhaps our elaborate nervous system provides us with assets unnecessary for life today but essential for survival tomorrow. Whatever the dreams we entertained in the past few centuries—during which humans advanced from bridling a horse to teleguiding a space probe—reality always outdistanced fiction.

So let's be beautiful dreamers: year 2050. Aftermath of world war. The sea and air, polluted; the continents in ruins. Ninety percent of plant and animal species have vanished, and many that remain are those that reproduce rapidly, like insects, and plague the rest. The human population, which had reached a peak of ten billion, has been reduced a thousandfold.

The beautiful dream? Our species survives.

We've finally learned the near-fatal lesson. We come together, abandon our rivalries. We discard the artificial concept of nation-states; we survivors now know we must cling to existence as a human community rather than tempt fate as opposing tribes.

One precious asset remains: Most of the past achievements in arts, science, and technology have been recorded on memory chips and buried in stainless-steel time capsules by prescient humanists who feared their future. These powerful tools are at last used constructively for the formidable task of the big cleanup—purifying water and air, planning rational progress.

As we delve deeper into our imagination and voyage farther into the future, we navigate oceans of time, traversing hundreds and thousands and millions of years. Now the 2050 tragedy is in the distant past, and human beings have long since abandoned research in the sciences of destruction. We concentrate all resources on influencing biological and geological phenomena in order to fend off Earth's own natural cataclysms. We know that if we do not intervene, the magnetic field will reverse; ice ages will come and go. In the slow process of continental drift, North and South America both will collide into Asia and re-create the Gondwana continent, but inversed. The rotation of Earth will slightly slow, then accelerate. If humans allow the moon to continue on its natural course, getting closer to us, the gravitational pulls will be so great on both its sides that the moon will tear apart; some of the resulting asteroids will bombard Earth, while others will form a belt around our planet much like the rings of Saturn. But, of course, we do not permit any of this to come to pass. Through the course of the past millennia we have used our growing knowledge to tame the forces that once, way back in the twenty-first century, were loose in the cosmos. Back then, humans had already toyed with nuclear energy and, remarkably enough at the time, had created the laser, a coordination of light that may not have existed anywhere else in the universe. Our monumental achievements today, in the second billennium A.D., now dwarf those sand castles. We have learned to master gravity; we can send Newton's apple back to the tree! To protect the moon, we modify its orbit; we control the ocean tides and climates. If need be, we can push stars apart or bring them together. We rein in many other forces as well, like antimatter and even energy from other universes.

Commanding the cosmos—however thrilling for a little man on a little planet—pales before our achievements influencing life itself. Humans have long fostered fantasies of retarding death. Even ancient Assyrian tablets, which recorded humankind's earliest dreams, tell the legend of Gilgamesh, the Sumerian hero who, in anguish at the tragic loss of his friend Enkidu, despairingly plunged into the sea to seek the herb of life eternal. Almost as soon as he found his prize of immortality, the symbolic serpent stole it away.

The secret to eternal youth is less elusive for us in the second billennium A.D. than to Gilgamesh in Assyrians' dreams. Until the twenty-first century, we were so accustomed to death that we did not stop to consider why we were surrounded by mortal beings. A species must be capable of adaptation. Birth and death—the succession of generations—have offered species facing environmental change an escape route: mutation. The shorter the lifespan, the faster the turnover, the more probable the escape. Flies, for example, had better odds than dinosaurs for successfully adjusting to change. Natural selection imposed its inexorable rule: dying for survival.

But the human being has been the great zoological eccentric. Because of the continual changes produced through the births and lives and deaths of legions of creatures, the human being developed a brain with which he could steer his own evolution. It was like learning to drive. At first, in the twentieth century, medical sciences attempted only to heal the sick, to save the weak; because those with defects survived, they could reproduce, and inborn deficiencies were often passed to offspring. In a sense we had inadvertently initiated a process of regressive evolution. Scientists devoted their best efforts toward reversing this trend. In short order they had disclosed secrets tightly locked in the tiniest components of the gene. In the twenty-first century they acquired the knack of repairing genetic flaws before conception, opening the future to a sturdy population.

We then waved that new biotechnical wand over the aging process. We slowed the growth of children to give them more time for education and then we suppressed aging altogether in adults. Senility and natural death, once grim inevitables that haunted humankind, were simply banished from our life program. Now, in the second billennium, we remain eternally young and free from disease, the utopian dream come true.

Protected from aging, protected from ailing, human beings are still vulnerable to mechanical accidents; individuals eventually die. We carefully monitor the planet's total human population, replacing only the casualties. Thus those who desire parenthood can still have children—once every ten thousand years or so. Bringing them up, of course, requires another fifteen-thousand-year commitment.

Now that a billion years have passed, we understand so much about the intricate machinery of life that we are in a position to repay our debt to our obscure evolutionary ancestors and begin to replace the species eradicated in the hecatomb of 2050. But why limit our gift of life to only the species that existed in 2050? Why, for that matter, restrict ourselves just to the creatures of which we've learned through fossil remains? Why not enrich the globe with all species, known and unknown, that ever existed?

How do we re-create unknown creatures? We launch Operation Lazarus, starting evolution over from its earliest beginnings. We accelerate the process through genetic engineering, compressing three billion years to about three million years; then we provide each mutant with an artificial surrounding appropriate for its survival, thus protecting it from natural elimination and protecting ourselves from elimination by them. We turn vast land and ocean areas into gigantic super zoos, where dinosaurs roam with millions of tentative species that heretofore had been unsuccessful. In the sea, the fabulous sixty-foot Megalodon shark, extinct in the Tertiary era, swims along archaic reefs, chasing jawless fish. Our super botanical gardens grow lush with such archaic vegetation as the scaled trees that thrived in the Carboniferous era. I imagine myself spending centuries in the botanical, zoological, and marine parks.

Of course, we wish to observe some of the odd creatures more closely. We remember with revulsion accounts of the unnatural and cruel prison-zoos of the barbaric twenty-first century. Those who would like to linger among the multitudes of beasts from Evolution II can stroll leisurely through holozoos, where three-dimensional holograms create the illusion of real jungles.

Maintaining our artificially inseminated planet requires a great deal of work, much of it in the oceans, obviously, where life proliferated most abundantly. To become more efficient within the sea, ocean-park wardens undergo simple surgery. Lungs and all bone cavities are filled with an incompressible neutral liquid; the nervous centers that normally trigger the breathing reflex are inhibited. A regenerating cartridge is inserted in place of the lungs in order to oxygenate blood and to remove carbon dioxide (a step initially taken in the twenty-first century to save lung-cancer patients). An entirely new brand of amphibious human beings enters the sea: *Homo aquaticus*, able to resist pressures down to five thousand feet, to descend to this great depth freely, then later to surge to the surface with no decompression problem at all.

Even *Homo aquaticus* needs the help of marine partners as deputy wardens. We choose orcas, with their capacity for intelligence, language, and longevity, three of the four essentials for civilization. We modify their genes to give them the only requisite they lack: handlike manipulators—retractable, of course, so as not to reduce their speed and agility in water. Because we get along so well with them, we endow them with immortality.

Our earth thrives, exuberant with luxurious gardens and populated by a controlled number of creatures in each of the millions of species. Our cousins the orcas have helped us develop an ocean ten times more rich in variety than the ocean of the primordial twenty-first century. We would not think of fouling such a world with pollution; we either annihilate, by antimatter, the waste products of our factories or send them to the sun.

At the dawn of the second billennium A.D., where do we find our own place in this new world? The ultimate civilization requires the ultimate in information exchange and storage. "Knowledge injections" do wonders for education as well as for research. Where handwritten script was once replaced by printing, and printing was later supplemented by the electronic media, those curious antiquities today are supplanted by our new biological brain wave transmitters, which we can switch on and off. Our ideas, experiences, and opinions, transmitted by brain waves, are picked up by satellites and instantaneously exchanged or even registered as votes. All decisions are made by the human and orca community. We no longer need leaders. "Democracy" now means more than was ever imagined by the ancient Athenians.

This mental communication system provides us all equal access to the latest technical developments. Also readily accessible are classical works of history, literature, music, painting, sculpture, architecture. Many of us enjoy the "synergy arts," a supreme combination of all creative disciplines. We don't purchase tickets—we close our eyes. Multidimensional reveries stimulate each art lover's senses of vision, hearing, taste, smell, and touch—daydreams created by our transcendent artists, in their ultimate performances, inspired by the full chorus of muses.

In the fields of science and technology, our immortal geniuses apply themselves to the search for ways to prevent our planet's incineration when the sun, as all stars do, inevitably expands in its spectacular and fiery throes of death.

They also work out elaborate schemes to bypass the upper limit set by the speed of light. Just the same, a now "old-fashioned" method of space travel, soaring on solar winds for thousands of years at half the speed of light, is perfectly pleasant for immortal voyagers, as long as they bring along interesting companions and equipment to produce tasty food. We use some of the desert planets of our sun as bases to fire salvos of the new superphotonic cosmobiles, crowded with immortal pioneers who will explore and prospect every last celestial body spinning in our galaxy. We sow human achievement throughout the Milky Way.

The inevitable question as we voyage in infinity and live forever: Is eternity boring? I don't think so. If eternity is boring, why would it have been proposed by all our religions as the supreme reward? Rephrase the question: Now that we are immortal, what has become of the basic motivations that nature has instilled within us? The concept of a territory, of vital space, is unimportant because our population is so small. Violence and greed? What for? We number only ten million on a spectacular and plentiful planet. If we still have greed and violence in our blood, we simply eradicate it. We learned in the old days to vaccinate against rabies; now we vaccinate against rabid behavior.

The other basic motivations, hunger and sex, present some puzzling questions. Can the pleasures of eating and loving become too familiar over the course of eternity? If our distant mortal human ancestors, even when they reached the age of eighty years, were never bored with food or sex, it was probably due to the fact that lingering sensations faded between experiences. In the second billennium A.D., if we want to increase our appetites artificially, we slightly reduce the memory of pleasure . . . so as to keep incentive strong.

Obviously, the human intrusion deep into the species' own nature—abolishing natural death, increasing physical and mental capabilities—raises crucial moral and social questions. Are they much more difficult to answer than those that twenty-first-century humans faced with abortion, euthanasia, gene manipulation? Not really. Moral rules are the rules of the life game. Now that we have broader perspectives on life, our new moral values are enriched. Commitment to the community has widened to commitment to humanity. Regard for the planet has expanded to respect for the cosmos. Our fundamental principle: Life is sacrosanct.

The human being now is moral, may live forever, has mastered the wild forces of the universe and penetrated the intimate sanctuaries of life. He knows the past and he can predict and even modify the future. He has finally reached the biblical ideal; he is created in the image of his Maker. Time as a dimension could be curved to close on itself, much like the surfaces studied in topology. God, creation, and the human become constantly present, eternal, inseparable.

Now, independently—in one or several corners of this universe or of another—other creatures have advanced as far as human beings have.

They become aware of some remote interference with their own plans to bridle the cosmic forces. They identify the resistance. They want to meet with their peers to avoid the catastrophes that could be unleashed if independently directed cosmic energies collide by accident. We convene to plan the peaceful management of universes. We hope, of course, that we—intrauniversal masters of matter, energy, space, and time—do not fall victim to the primitive passions, jealousies, rivalries, and hatred that plagued our primitive ancestors of the twenty-first century . . .

* * *

. . . Day has dawned. The sun, its full round sphere burning bright, climbs in the sky above *Calypso*. I shake myself from my reverie. A crew member arrives on the bridge to relieve me of my watch. He hands me a fax that has arrived from my Paris office, several time zones ahead. It says that the nth Disarmament Conference has once again been scuttled by a few nations that cannot agree on a five-year plan.

We have such a long way to go.

CHAPTER ELEVEN

THE MIRACLE OF LIFE: THE HUMAN, THE ORCHID, AND THE OCTOPUS

MY friend Albert Falco and I glided at a leisurely pace over the floor of the Aegean. The strain of our expedition was beginning to wear us thin. For weeks we had been excavating the sunken remains of a Roman galley only a cable-length away from the steep rocky shoreline of the island of Antikythera. Loaded with spoils from the pillaged city of Pergamon, the ship had been swallowed by the avenging sea some two thousand years ago. We were paying dearly to salvage the plunder: diving twice a day to difficult depths; clinging endlessly to *Calypso*'s keel ten feet below to decompress; hauling tons of sand and silt while knowing, as our muscles ached, that just one slip could destroy some priceless muddied treasure.

I could feel my spirits revive as we slid aimlessly through the deep, hovering from time to time over clusters of sponges fastened to the seafloor. Commercial divers had long since stripped the bottom of the lovelier marketable sponges; only common species of no monetary value remained. The bowl-shaped, dark-brown sponges lay strewn across the floor as far as the eye could see.

We coasted over the mouthless, organless, motionless forms. Suddenly something caught my eye. Deep within one of the sponge bowls lay a half-inch heap of sharp-edged, white debris that resembled broken porcelain. I looked around. Each and every sponge within view cupped a neat pile of smithereens.

We surfaced and pulled off our masks. What on earth, I asked Falco, was going on? Days passed; the solution to the puzzle eluded us.

Finally we solved our mystery. The rocky crevices below were lavishly adorned with bryozoans—beautiful, brittle, coral-like colonies that twist like lengths of crystallized crochet and shatter at the touch. Small fish called wrasses were breaking off pieces, carrying them to the sponges, and then dropping the shards into the sponge bowls to hide the eggs they had laid there!

This was breathtaking behavior for little fishes. Why were they doing it? I thought of the time we wanted to show that sponges feed by constantly pumping quarts of nutrient-rich waters through their filters. We had injected dye into the bottom of a sponge living on the seafloor; almost instantly, a column of colored dye jetted upward out of the top of each sponge. Of course! Here was the explanation for our wrasses and their sponge-bowl nurseries, at once obvious and almost unbelievable: The wrasses had instinctively laid their eggs in a place where they could be oxygenated with fresh supplies of water while still being completely hidden by the broken shards of bryozoans.

Time after time I have been in the sea, or on its surface, or at its edge, thinking that I was witnessing the supreme example of nature's intricacy—a miracle to surpass all miracles. And time after time, the natural world has then transcended itself, offering up some new incredibility that once again has left me speechless. The second law of thermodynamics, defining entropy, dictates that the energy of the universe is relentlessly degrading, that the cosmos itself is slowing toward an end of total inert disorder. Yet all around me life seems to have defied this universal law, constantly becoming more elaborate, creating highly complicated organic molecules, organizing chaotic matter into structures of trillions of cells—like my children and grandchildren. Father Teilhard de Chardin, contemplating Earth's phenomena, once wrote of a third infinity. There is the infinitely large, of course, and the infinitely small. To them Teilhard added the infinitely complex: life itself. The miracle of miracles, finally, is that life exists.

It is not just poetically apt that the sea should have provided me with abundant examples of life's wonders; it is geologically inevitable. The most numerous and diverse forms of life inhabit water because life began in water. Somehow life took shape from organic molecules in the sea. The first monocellular algae, through photosynthesis, produced oxygen. Perhaps to protect themselves from the corrosive effects of this gas, cells developed membranes; sexual reproduction, variety, and abundance followed. Seen one by one, each step seems a laborious, hairbreadth victory, a lucky accident. But together, they are an orchestration of accidents. At any moment a clashing note, echoing out of the universe's infinite skies, could have stopped the music. But instead of vanishing in some cosmic cataclysm, our planet survived. Washed with liquid water, a compound on which life depends, life on Earth continues to proliferate, to elaborate. In its relatively short existence, life has evolved into the forms biologists consider the most complex vertebrate, the human being, and the most complex plant, the orchid. I would designate the octopus, given its intelligence and devotion to the continuation of its species, as the top invertebrate. The resounding chord in evolution's symphony: the human, the orchid, and the octopus.

The evolutionary trends that began billennia ago continue to ramify and repeat themselves with each new birth on each new day. On board *Calypso* we haul up plankton nets and occasionally find eggs caught in the meshes. They appear, to the naked eye, to be identical. Yet one will turn to sardine, one to mackerel. Which? I slide a bigger egg under the ship's microscope. Through the transparent capsule, shining jewels of color focus into the pulsing shape of a tiny squid; in a few days its bulges will metamorphose into eyes and organs. How? By the time the hatchling has grown to juvenile and then to adult, the vital data for nurturing its own eggs will have been imprinted. When? If the egg and sperm carry information only for faithful duplication, if experience and learning are not transmissible through genes, how, then, is the wrasse instilled with instructions to hide its eggs in a sponge bowl that will provide the generating offspring with oxygenated water? For that matter, why did the trigger fish we watched in the Red Sea automatically swell its cheeks and blow water across its own eggs? Why did the octopus we filmed in the Mediterranean intuitively squirt her hatching embryos with water from her funnel? Why and how and where is ancestral experience programmed into hatching eggs?

Lewis Thomas, the American physician-philosopher, once described the quizzical delight felt by all observers of this miracle of life:

> You start out as a cell; this divides into two, then four, then eight, and at a certain stage there emerges a single cell which will have as its progeny the human brain. People ought to be walking around all day, talking of nothing except that cell. If anyone succeeds in explaining it, within my lifetime, I will charter a skywriting airplane, maybe a whole fleet of them, and send them aloft to write one great exclamation point after another, until all my money runs out.

I know just how he felt. No number of exclamation points could possibly describe my sense of wonder when, skin against skin with water, I've not only witnessed life's miracles but mingled with them: holding my breath to hear the distant songs of humpbacks; opening my lips to let salt awaken my ancestral taste; gliding with banks of tuna, my muscles beating in unison with theirs. Albert I of Monaco once wrote that "after watching the sea through calm and storm, heat and cold, day and night, one comes to hear and understand her language."

And what wondrous tales the sea can tell. Each time I've opened my eyes below the surface of the sea, I've noticed something new; I've stolen glimpses of creation. Scattered Cassiopeia, a kind of jellyfish, lie on their backs, letting light vitalize the personal algae gardens they carry on their undersides, steadily undulating as ripples of water wash their growing vegetables with nutrients. Appearing almost to merge with the sea, the transparent, barrel-shaped salp's body is punctuated only by a minute orange dot at its center. Hundreds of these crystalline creatures link together, creating "Venus girdles"—long, silvery, diaphanous belts that waft into spirals, modeled by the slightest water eddy, formed by individual animals so glassy they seem to be water come alive.

With life itself having been born in the sea, perhaps all nature's complexities fall under that simple rubric—"water come alive," a miracle that embellishes all the miracles I've seen.

○ One Antarctic night, as the low-hanging sun drenched *Calypso* with an unreal pink light, orca coughed blows of vapor a cable length from the ship. Below our keel, swarming marine life appeared more plentiful than in any tropic I had seen. Deep ocean currents had swept rich organic matter from all corners of the sea and upwelled them to the surface of the Antarctic, where in the uninterrupted summer light, they had turned icy expanses into glacial meadows. Improbable "ice fish" grazed in waters only a few degrees from freezing solid. Red-faced from the brutal cold, I watched life teem at the edge of death.

○ Often, on the ocean floor, we have come upon tiny holes in the sand surrounded by unidentifiable animal tracks. The bottom plains stretch for miles, providing no natural shelter. Are the holes, I wondered, entries into secret hiding places? I had an idea: We poured a fast-setting plastic into the openings. We let it solidify, and then pulled out an artwork. The molded plastic looked strikingly like an upside-down tree. Burrows, forks, coiling tunnels, and escape routes: Under each square foot of sea bottom branched several feet of excavations. Even the sediment is populated; out of sight, underground, prey and predator perform their roles in the drama of life.

○ As we were sailing from Toulon to Greece, we gathered on deck one calm, moonless night to watch what appeared to be fireworks shooting out of the glassy sea. Innumerable little flames rocketed from the inky water, then fell back against the surface with the soft murmur of raindrops. No fish we knew played games like this. Years passed before we realized that our entertainers had been jumping squid with luminescent photophores. A scientist on board *Calypso* once speculated that the creatures might use their lights as schooling beacons in the dark or as glowing lures for gullible prey. As yet, no one knows for certain why squid whip up stardust storms that rain at sea.

○ Tantalized by fascinating reports from scientists, we sailed *Calypso* to the Yucatan Channel. There, I waited for a northwest winter wind, then organized an underwater night patrol: I wanted to be sure we wouldn't miss the amazing lobster march that the scientists had described. Once a year, they had told us, always after the first winter storm and always at night, the arthropods parade. We were right on time: Below, we saw thousands of lobsters, each reaching toward another with its antennae, assembling into groups. Suddenly, the leaders all moved out, as if in response to a silent gunshot to "be off," with queues of lobsters following behind. If challenged by our lights, the lobsters would circle like besieged wagon trains; as soon as the threat was withdrawn, onward the perseverant creatures marched. Even predators could hardly disrupt the file; a single lobster would step out of line to defend his fellows, most frequently a self-sacrifice that gave the others time enough to forge ahead. More than a million lobsters, by my guess, joined those ranks, moving almost a mile an hour, soon disappearing too deep and too far to be followed, heading for a destination no one has located for reasons no one has fathomed.

We've witnessed ample evidence of Teilhard's infinitely complex; yet we've watched, too, as complex creatures all conform to the same patterns of the earth and sea, obeying the silent natural commands that unite all life. The cycle of day and night, for one, seems to pull the strings of life like a puppeteer. Off Madeira Island, for example, the bronze-black espada fish, extraordinarily sensitive to light, react precisely and predictably to the faintest gleam from above. Fishermen can catch them only at night in very deep water. Under starlight alone, fishermen set their hooks and lines at a depth of 5,000 feet; when the moon shines as well, they must lower their bait even farther, all the way down to 7,000 feet. Fishermen have sworn to me that they must raise or lower their gear 1,000 or 2,000 feet as the moon slips in and out of clouds. In waters two times too deep for the most brilliant sunlight to penetrate, the espada somehow obey the glowing moon and flickering constellations. Beyond the narrow margins of human understanding, only slightly widened by technology, we sense nothing and comprehend little.

Nature's day-night cycles command the most pervasive obedience from the sea's far-reaching Deep Scattering Layer. In our early years on

Calypso, we noticed, each twilight, that our echo sounders would routinely indicate what looked like a false bottom smoothly rising toward the surface. Each morning, just before dawn, the apparition descended. Our friend Doc Edgerton lowered camera equipment to photograph the phenomenon, and the resulting photos suggested that the rising layers are composed of trillions of unidentified light-sensitive creatures.

When I descended into deep waters during France's tests of its bathyscaphe, I didn't count on passing through the layer; the ocean is immense, the bathyscaphe nothing more than a needle prick . . . unless the Deep Scattering Layer was as extensive and teeming with life as we believed. After all, I thought, behemoth whales can't live on water and love alone.

Eighty feet down, we saw the first clues that supported our hunches about this magically ascending and descending false bottom: foggy whirlwinds of plankton, a few little fish, and minute crustaceans. At 165 feet, creatures congregated even more intensely, as hundreds of small squid swam upward, glowing like gas lamps on a murky London night. At 330 feet, we were still enveloped in the soup, within which tiny shrimp headed up toward the distant surface. The nightly migration, inexorably rising and falling, was triggered by light—generator of life through photosynthesis, manufacturer of oxygen, architect of beauty.

We have, on our voyages, witnessed an even stronger tie than the natural bond that joins creatures to the sun, moon, and stars. All animals seem to share the same motivations. Warm-blooded and cold-, swimming and striding—all beings are driven to fill identical needs: for food, for territory, for individual survival, and for the survival of their species.

The infinite variations in the ways creatures fulfill the same requirement—to fuel energy needs—constantly astound me. Booby birds and pelicans were once thought to feed by careening, from above, down onto fish, plunging into water only because of the speed and force of their descent. On an expedition to Isabella Island, however, I watched from within the sea as these birds actually performed underwater dives, descending some twenty feet below the surface and then flapping their wings to fly through water. Totally encrusted with tiny diamond bubbles—like the jeweled nightingales of Asian emperors—they soared around below for nearly half a minute.

The sea otters with which we've played in California, as mammals that once lived on land, still retain digits and can manipulate tools. I've watched these elfin creatures use a stone to crack open their favorite delicacies—sea urchins, king crabs, and abalone. Sometimes, a sea otter floats on his back and places the stone on his stomach, using the stone as an anvil against which he crushes his appetizer open as though he were an expert epicure.

Civilized humans—carnivorous themselves—never seem to adjust to the routine reality that a carnivore's meal is a bloody one. I have seen sea lions catch sunfish, tear off all the fins, and store the living but helplessly crippled creatures in some crevice within the sea, to which the animals may return when they are hungry. If a human takes the point of view of the sunfish, he would, I suppose, regard the stripping and storage process as cruel. From the sea lion's perspective, though, the behavior seems ingenious. Like humans who choose lobsters from tanks in expensive restaurants, the sea lions have learned to keep food fresh.

Some insects have developed an even more "inhuman" larder. The ichneumon wasp supplies its young with all the nourishment they need to sustain themselves to adulthood. The wasp digs its eggs into a caterpillar. There the larvae live, eating just enough to spare the life of their host. Some might call this ruthless. Is it really? According to whose morality? God's? The wasp's? The creature is, after all, simply ensuring the life of its progeny.

The seeming confusion that broils in the sea as creatures vie for living space, hunting territory, and sexual partners also neatly falls into a single natural order: the drive for ownership. The tenacity with which some sea creatures cling to the shelters they have staked out for themselves conjures up hapless little homeowners. Coral reefs abound in examples. Almost each animal has his special home: the damselfish, the ramified coral head; the surgeon fish, his slot; the clownfish, the protection of venomous anemone; the bulldozer shrimp, his deep, tiny hole in the bottom, guarded, of course, by his doorkeeper, the goby; the lobster, a shipwreck or an alien piece of pipe. The butcher, the baker, the candlestick maker: The sea is full of bourgeoisie—timid, sedentary creatures that keep their homes tidy and lock up each night.

Maybe as a nomad at heart, I've found my own admiration captivated instead by the creatures that seize their property in spine-tingling ways—the migrators. Traditionally, naturalists have said that these transient animals claim no territory. To me, they seem to claim all territories. A raider like the shark, with his acute sense of smell, keen vision, and nerve-flooded flanks, registers every intrusion into the circle of water he commandeers as his own. When we've trespassed into those unmarked but unquestionably bounded areas, the whole sleek mass of snout and teeth and flesh jerks at the first ripple of our pressure waves. The warning speaks for itself: My territory is where I am, boys.

Dolphins stake out the high seas as their own, cutting along their migratory paths like bullets. In any litany of miracles, the dolphin's speed deserves priority placement. Once, when the Navy had assigned me to service in the Far East, I was standing on the foredeck of the *Primauguet* during speed tests in the Indian Ocean. When we hit thirty knots and I had to bend into the wind to stay on my feet, I watched in disbelief as a group of dolphins approached us from behind, swam effortlessly alongside, then played in front of us for a few minutes before they dropped away. Amazed by the fast tempo of their undulations, I had counted their tail strokes. The cadence almost matched that of our propellers. To me, the animals seemed to be merely measuring their own mastery.

As gentle as dolphins are, their young are hellions in the ways they wield their claim to what I like to call fluid territory. They storm the sea like unruly teenagers, the punks of the marine world. In the Indian Ocean, we watched a pod of young dolphins plundering everything, eating everything, frightening everything, wildly and savagely breaking all the zoological rules (wild creatures almost never otherwise squander food). The restless energies of these youth demand to be spent; wrack, ruin, waste are their instruments of catharsis. Young tuna, squid, the cod that lay siege to the bays of Newfoundland—they all charge like Mongolian Huns into quiet little seafloor villages, marauding and ransacking the abodes of all the peasantry, which scamper to their peepholes and emerge to pick up the debris when the ocean's underaged Attilas have swept on. I imagine that during the Stone Age, five- and six-year-old human children pillaged the lands in similar gangs, attacking adults in uncontrolled outbursts nearly identical to those of the sea's enfants terribles.

The drive for territory feeds violence; and nature's violence itself is unarguably another of the world's wonders. Although groupers have behaved around us as if they were personal pets, they fiercely prohibit neighbor groupers from crossing their property lines without their tacit approval. Once, to see what would happen when this unwritten rule was broken, we surrounded a grouper's den with four mirrors. The fish was befuddled—but soon gathered his senses and charged his reflected intruders, shattering the mirrors and then angrily biting our diver as well.

We learned that even fixed animals like coral can fight as aggressively in their competition for space. With time-lapse photography, we filmed coral actually throwing out filaments that spewed deadly digestive acids over a less-assertive coral species nearby. Over the hours, the neighbor succumbed, corroded alive. Finally all that remained was a skeleton; even that soon was invaded and possessed by the enveloping tissues of the victor.

Male elephant seals also engage in combat, biting through flesh and heaving their corpulent three-ton bodies at contenders for their harems, though they usually end their skirmishes short of killing. Although I've watched and filmed the elephant seals demonstrating their strong sense of territory for mating, I was nonetheless startled to discover that they claim a territory even for death.

We were studying the elephant seals in Guadeloupe when my son Philippe and his camera team decided to explore the waters around the island. The men were distracted by the sight of a floating carcass—a dead baby elephant seal, probably drowned while learning to swim.

The legend of an elephant graveyard had long echoed across the globe. In darkened saloons throughout Africa, those who traffic in white gold whispered rumors about a cache of skeletons—and ivory—in some shadowy corner of the wild where the elephants went to die. I can't be certain, of course, that any of my team members had heard the tales. Perhaps even Philippe and the men didn't know just why they began their descent under the floating omen at the surface.

They spiraled down the barren underwater terraces of Guadeloupe's volcano. One hundred fifty feet below, by the light of their underwater torches, they laid eyes on a ghostly world—a field of skeletons, whitened bones of creatures that had not been killed and disposed of, but that, once dying, had themselves all obeyed some silent call to the same site. I wondered at first whether a watery catacomb of unburied bones was really an example of animal territoriality; but the tomb was, after all, a property chosen not by the dead but as the last conscious act of the living.

If humans, too, instinctively obey the drive for territory, they have consciously endorsed another drive—for survival—by condoning even violence as justified in "self-defense." In the sea, the drive to survive takes a multitude of forms.

The simplest means of self-defense are armor, like the sea turtle's shell; shelter, like the burrows of garden eels; and escape, either by bursts of speed, like the flight of the octopus, or by seemingly magical capacities for camouflage, like the élan that a flounder displayed in developing black and white squares when we placed him on a checkerboard. We have, however, been astonished by nature's far more complicated defense strategies. Occasionally, in tropical waters, I have come across glittering silver walls, rising all the way from seafloor to surface. Tens of thousands of sparkling grunts have fashioned themselves into a Japanese screen of shimmying, shimmering living beings. Each fish in the school is girded with lateral lines that are sensitive to pressure; together they can flee danger with uncanny synchronized precision, navigating their dazzling wall like a single creature with a single brain. They break from one swirling "organism" into two when necessary, confusing marine predators that, unlike land animals, have enough difficulty as it is pursuing dinner through water's three dimensions.

Near the Shab Rumi reefs, in the Red Sea, I have seen the outrageously colorful parrot fish spend half an hour patiently weaving himself into a transparent, glassy cocoon, which he spins from saliva and attaches to the crevices of coral reefs. There he rests, a sleeping beauty cradled in bubbles of cellophane, lovely to behold but poisonous to any animal that might take a bite.

If we hadn't filmed the ballet of the sea scallop, I might still think it was the product of an animator's imagination. Throughout the day, scallops lie scattered across the seafloor looking like so many fossilized lady's fans. Yet when scallops recognize their specific adversaries, starfish—and they can, because they have numerous eyes and use chemical receptors to sense the enemy's approach from behind—the shells all rise. They stand on end, opening and closing to create water jets, and they take off, spasmodically pumping their jets as they comically dance away.

From our diving saucer, Falco and I have often seen defense take the form of deceit. Fleeing squid leave behind clouds of white ink as decoys to mislead predators. The most startling stratagem for self-defense: vanishing from the sea altogether. The fish known as exocets abound in warm open oceans and provide daily manna for the speedy mahimahi. To elude the mahimahi onslaught, exocets fly. They leave the water, spread their winglike fins in the air, and glide in the wind for a half minute. What amazed me, when I observed this life-and-death battle of abilities, was that while a mahimahi continued to charge forward for the purpose of catching the exocet at its splash landing, the exocet further deceived its pursuant by changing course in midair, then plunging down, into at least momentary safety, in an unexpected landing site.

All three motivations—eating, owning, defending—would be meaningless without a fourth. In nature, even survival of the individual is subordinate to survival of the species, on which life—not one life, but all life—is ultimately dependent. Creatures are survival machines, programmed to ensure the bequest of their DNA, the perpetuation of their line. This sacrosanct élan vital, passed from one generation to the next, endows procreation with beauty as the ultimate act of life.

I can think of countless ways that nature ornaments her unions. The Adelaide penguins, which we studied at Deception Island in Antarctica, are the most loyal and affectionate birds I have ever met. Penguins are monogamous, remaining faithful to their partners from one season to the next. Even though they spend the long winters separated, feeding in the sea and traveling along the Antarctic continent, when the mating season approaches, they return to the very rookery where they were born. The males generally arrive first. Sitting and stretching their necks in cadence, they vocalize wildly to attract their mates. Despite the eight-month separation, each female unerringly identifies her partner by his voice. The courtship rituals continue, with the males offering repeated gifts of pebbles to the females, who build nests of stones. The gentle pairings are followed by cheek-to-cheek demonstrations of what would appear to any human onlooker to be tenderness. When their eggs have hatched, the parents courageously fight predators away from their offspring. Parent penguins raise their chicks with strict discipline and apparent love. Then, having passed the baton in the relay of life, they plunge off again for another long, adventurous separation.

We have also filmed thousands of groupers instinctively heeding the irresistable call to crowd into the tiny coral gardens of Caye Glory, near Belize, where they ensure the continuation of their species through their sheer numbers. Groupers are embellished hermaphrodites; born female, they turn male at maturity. Couples spiral frenetically, their colors changing as they spawn. Males darken from orange to dark brown; females turn pale, and their stripes vanish. Then the shades fade and brighten in reverse, as the groupers form a carousel of blushing hues while one gives milt and the other, eggs.

The behemoth right whales of Patagonia also celebrate awesome nuptials. We have filmed two males pursuing the same female, who at first turned her piebald belly to the surface to avoid her two suitors below. They seduced their reluctant Dulcinea by gently stroking her with their giant fins. When she finally rolled in submission, the mammoth bulls shoved each other in competition, each of their penises extending perhaps five feet, until one victorious male won her favor. After only eight seconds, the first bull disengaged to rise for air, while the second took advantage of his absence. The coupling lasted for hours, "on the blue deep bed of the sea," as D. H. Lawrence wrote, "mountain pressing against mountain, the zest of life."

As breathtaking as witnessing such unions can be, the moments that have most inspired me, when I've been privileged to see them, are those in which creatures sacrifice their own well-being for their offspring, when they even give up their lives in the act of giving life, protecting their species before they protect themselves. The tilapia fish incubates her eggs by gently holding them within her large, scooplike mouth, carefully keeping her jaw open while she brings them to term. Even when her young have hatched, and venture into open waters, they rush back to the safety of their mother's mouth at each sign of danger. We also filmed the five-week incubation of the octopus: The mother glued her eggs to the ceiling of her cave, where they hung in clusters like ripening grapes on an arbor. Then she settled at the entrance, eight arms outspread. Periodically vacuum-cleaning her eggs with her suction cups, she refused to eat any of the food offered her—to the point that she would die soon after her eggs had hatched. She would not permit food to pollute her all-important eggs, her promise to the future.

We watched the sea-arrow squid—the sparkling creatures that once rained like falling stars around our ship—in what seemed to be almost a festival of self-sacrifice. As the excitement of the males intensified, their flesh darkened to maroon, and they vehemently and urgently embraced the females. Inseminated, the females labored to extrude their strands of ova, continuing, in exhaustion, to incite the males to fertilize the eggs even after they'd been laid. After twenty-four hours, starving males paused in their compulsive mating attacks only long enough to cannibalize the weakened females; other couples, sapped by the creative process, died still hammer-locked together in tangled arms. Just five days after the maelstrom of mating began, our divers saw a deathly calm: Twenty million tiny sea-arrow corpses lay scattered among the billions of throbbing, diversifying, living eggs below.

During our expedition down the Nile, we witnessed the most memorable cost that nature extracts as the price of life. We were flying low, my son Philippe and I, in our Catalina seaplane. Below us, dense tropical forests blanketed the islands of Lake Victoria, in the heart of the African continent. The scenery was at once monotonous and beautiful.

After an hour, the eastern horizon caught our attention. A series of straight, dark columns, funneling upward to the sky, proceeded in a line across the lake. Were they tornadoes? I asked Philippe to proceed with caution. It took us twenty minutes to draw near; once we came close, the mystery only deepened. At least a dozen of the tornadoes were almost perfectly aligned in the same vertical plane and—could this possibly be true?—they didn't appear to aspire any water from the lake.

Suddenly, the windshield of the Catalina went opaque, thick with little black spots. I say with some wry humor that the viewports were fortunately open, and so our mystery was quickly solved. Millions of tiny flies invaded the plane. The "tornadoes" were thousand-feet-high columns of insects! They rose in teeming spirals from the lake. Individually, each fly was innocuous; together they could easily stifle an airplane motor.

We landed the Catalina and steered our Zodiac into the insidious buzzing masses. The scientists advising us for the film informed us that once a year the notorious Lake Victoria flies mate and then die, leaving their larvae to incubate in the lake waters. During the twenty-four-hour mating season, we watched each column throb with more than a billion creatures, an inconceivable figure under normal circumstances, but there, engulfed by a billion insects, I felt I had a concrete notion of the number. Though we protected our faces with mosquito nets, the tiny bugs got in our eyes, our noses. Too small to be well recorded by our camera, the flies on film would look like clouds of dust kicked from the wheels of a truck trundling through the desert.

The next day, as we flew to Entebbe, I saw another regiment of columns. The lake had given birth not to a billion, or several billion, but to *hundreds* of billions of flies, each one dispensing thousands of eggs. The wind had pushed some of the insect clouds off the lake and onto land, where even the trees were "smoking" with rising fumes of insects.

I thought to myself: Here was the act of life stripped to its fundamentals, like a diagram. The eggs would incubate for a year in obscure development, in their slow metamorphosis in the dolorous hazards of the lake—all for one single apotheosis, a whirlwind of life that would end almost immediately. Because of the crazed squandering of their vital energies, the species as a whole survived—not in spite of, but due to, this fecundity of death.

I thought of the gasoline that powered the motors of our airplane, petroleum created from the remains of myriad tiny creatures entombed for eons by geological upheavals of the planet. The truth struck me: To observe the dramatic saga of the lake flies, we were burning the fluids of existence.

Each year, in spring, swarms of lake flies procreate and die in their life-giving hecatomb. This mix of frenzy and death, this inexorable alternation that fuels life, can be readily seen in creatures of brief existence. Is it possible that when two human beings join to create a child they retain—buried somewhere in their primitive subconscious—the memory of an act that, at the Origin, levied the ultimate price of death in exchange for the supreme gift of life?

EPILOGUE: AN UPDATE SINCE THE WRITING OF THIS BOOK

By Susan Schiefelbein

JACQUES-YVES Cousteau and I finished this book shortly before his death in 1997. He and I had been collaborating on various writing projects for more than twenty years—a period during which, always with an eye on writing a book, we gathered the materials that appear in the text. During those years, Cousteau issued many warnings about the directions in which human beings were headed. Whenever confronted with the prospect of calamity, he searched for ways out. His foresight proved to be exceptionally clear. Some of the exit doors he identified have, in the intervening years, unfortunately slammed shut. Most of his solutions, however, not only remain valid but require urgent attention. Herewith, a brief update on events that have transpired in the ten years since his death.

CHAPTER 6: SACCAGE

In the late sixties, Cousteau began to notice a marked diminution in the numbers of creatures he'd seen at the beginning of his career some twenty years beforehand. He decried not just pollution but also what he called "saccage," destructive human practices, and he called for a "vitality index"—some kind of way of measuring what he felt certain was the resulting disappearance of life forms.

○ Since that time, 20 percent of the world's mangrove swamps, which are both rich ecosystems and a vital buffer between coastal communities and the ocean, have disappeared.

○ Despite the fact that half of all plant and animal species depend on tropical forests—and that half our medicines come from nature—50 percent of the world's forests have already disappeared. Although deforestation has slowed somewhat in North America, Europe, and eastern Asia, according to a 2007 U.N. report, worldwide we are still losing an estimated thirty-two million acres of forest a year, an expanse roughly equivalent to the size of Greece.

○ Cousteau repeatedly emphasized the importance of undersea meadows of Posidonia, the grasses in which fish lay their eggs. He feared that both saccage and pollution, which wreak their havoc close to the coastlines, threatened these "nurseries of the sea," which also protect against beach erosion. *Posidonia oceanica*, the variety native to the Mediterranean, is now an endangered species.

○ At the time of Cousteau's death, experts estimated that some 20 percent of all species on Earth were in danger of disappearing by the turn of the twentieth century. In fact, in 2006 the United Nations issued a ninety-two-page "Global Biodiversity Outlook" study, reporting that "we are currently responsible for the sixth major extinction event in the history of Earth, and the greatest since the dinosaurs disappeared, sixty-five million years ago." The causes—including overexploitation and climate change—show no signs of abating. Still, the report concludes that if we make unprecedented efforts, we may be able to slow the rate of extinctions by 2010.

CHAPTER 7: CATCH AS CATCH CAN

In the mid-eighties, Cousteau expressed disbelief regarding the past decades' worth of promotional materials published by world fishing organizations that touted the seas as an inexhaustible source of protein for the poor. The world's catch was increasing incrementally, he thought, because the world's fishing fleets were increasing exponentially. He asked me to perform what we both thought would be a simple task: to gather statistics comparing the increase in catch with the increased efforts to obtain it. I soon discovered that no world organizations published the figures necessary for comparisons. An officer of one agency gave me his home telephone number. Privately, he then confessed to Cousteau that agencies withheld statistics regarding increasing fleets because they risked having member nations quit if the true story became public. Cousteau and I set ourselves to the task of gathering figures on our own by researching individual fishery collapses. Today, twenty years later, the Food and Agricultural Organization of the United Nations publishes fleet statistics. Many fisheries, however, have already been lost.

- About 20 percent of the ocean's fisheries have collapsed.

- Since 1989, there have been no increases in global catch despite enormous increases in sophisticated technology, which allow even the tiniest of fishing boats to target catch down to thousands of feet with sonar. The World Health Organization reports that "most of the world's fishing areas have reached their maximum potential, with the majority of stocks being fully exploited."

- In the past, the closure of a fishing ground would bring back species whose populations had not been diminished by more than 90 percent. Populations of North Atlantic cod, however, have fallen by 99 percent, and though the fishery has been closed for years, stocks have not recovered.

- About 90 percent of the ocean's carnivorous fish, including cod and tuna, have been fished out of existence.

- Aquaculture, according to Dr. François Sarano, who for thirteen years served as Cousteau's science counselor, is nothing less than a "siege" on the poor. In their sea of statistics, international organizations praise the potential of aquaculture. Compared with the eighty-five million tons of fish fishermen catch each year, aquatic farmers produce sixty million tons, a seemingly impressive figure. These sixty million tons, however, do not substitute for lost ocean fisheries. If the tons of algae, mollusks, and fresh-water fish are subtracted from the statistics, aquaculture can be seen to produce only six million tons of seafish and crustaceans. Worse, the fact that the international corporations engaged in aquaculture are raising carnivorous fish gives birth to a scandalous figure. In order to produce one pound of luxury farmed fish, they must provide five pounds of live feed—namely, the mackerel and anchovies on which the undernourished poor, who depend on fish for food, have survived.

○ Fish consumption has increased in industrialized nations where the wealthy can afford such luxury foods as sushi. Per capita consumption of some sixty-one pounds annually, however, provides only about 8 percent of the protein eaten by these overindulged populations. Elsewhere in the world, however, about a billion people depend on fish for survival, and their future is threatened by what the United Nations calls "potential depletion of this important marine source of nutritious food."

○ Fishery agencies still have no means of enforcing international accords.

CHAPTER 8: SCIENCE AND HUMAN VALUES

In the mid-eighties, Cousteau lamented the fact that the United States, long a leader in funding pure science, had for some two decades been funneling financial support toward applied science and technology to the detriment of pure science's exploration and discovery. He feared that other nations would follow suit.

○ Between 2001 and 2004 the U.S. Defense Advanced Research Projects Agency, which had long funded pure science, cut its grants for academic-information research by 50 percent with the rationale that because the country had entered the 2003 Iraq War, scientists needed to focus on the shorter term. The policy makers failed to consider the fact that during the years of the Vietnam War, long-term basic science funded by the Defense Department ultimately produced what would become the underlying building blocks of the Internet.

○ In 2006, C. H. Llewellyn Smith, the former director general of CERN, the European Organization for Nuclear Research, commented on the funding policy of almost all countries belonging to the Organisation for Economic Cooperation and Development. "Now . . . governments will invest in basic research only if it can be shown that it is likely to generate rather direct and specific benefits in the form of wealth creation and improvements of the quality of life. This is a bad policy . . . and may actually be economically counterproductive."

○ Also in 2006, the directors of Germany's leading research organizations wrote an open letter to their country's minister of federal research, calling on him to reverse his policy of channeling funds away from pure science to research targeted on economic priorities. The minister ignored them.

CHAPTER 9: THE HOT PEACE: NUCLEAR WEAPONS AND NUCLEAR ENERGY

When we finished this chapter in 1996, Cousteau had already been protesting for decades against producing nuclear energy in a world in which its by-products, highly enriched uranium and plutonium, can be used for building nuclear bombs. He worried about the growing gap between the wealthy and the impoverished; he worried about the prestige attached to becoming a nuclear-armed nation; he worried that the International Atomic Energy Agency (IAEA) had not been endowed with the power to carry out its mandate. He foresaw that television and the

Internet would become incendiary ingredients when added to the mix of poverty and religious fanaticism: The disenfranchised would in the course of a single generation gain full information about everything of which they'd been deprived. He feared that our proliferation of fissionable materials could end in terror and catastrophe.

- It is tempting to feel optimistic about the Nobel Peace Prize awarded, in 2005, to the International Atomic Energy Agency and its director general, Mohamed ElBaradei, for "efforts to prevent nuclear energy from being used for military purposes and to ensure that nuclear energy for peaceful purposes is used in the safest possible way." Nonetheless, between 1997 and 2005, the IAEA failed to uncover secret nuclear programs in three countries.

- In 1996, experts predicted that the twenty-first century would open in a world in which nations aim some 20,000 warheads at each other. The actual figure in 2000 was 20,851. By 2005, the world's arsenals contained more than 28,000 nuclear devices.

- In 1959, Cousteau listened during an IAEA conference as technologists dismissed the fact that no long-term storage method had been conceived for radioactive waste. They blithely urged the world to "have faith" in the future. As this chapter was being written in the mid-nineties, the U.S. National Academy of Sciences was declaring that the identification of a storage method required "urgent attention." Now another ten years have passed. The *Bulletin of the Atomic Scientists*, founded after World War II by the physicists of the Manhattan Project, reported in early 2007 that nuclear-waste buildup is a "growing concern." "No long-term solution is in sight for disposal of nuclear waste," the report said, and "many plants are storing three times as much as their temporary storage pools were originally expected to hold."

- While nuclear reactors are fortified against terrorist attacks, the storage pools containing radioactive waste are not.

- Although the world still has conceived no long-term storage method for nuclear waste, since 1996, the nuclear industry has continued to produce an estimated annual ten tons of unirradiated plutonium (which has the most dangerous implications for proliferation). Exact figures for worldwide stores of plutonium remain unknown, as few countries submit annual reports regarding their stockpiles, and even those countries turn in incomplete or ambiguous data.

- In 1996, Israel, India, and Pakistan were suspected of stocking 90 to 275 nuclear bombs, with North Korea suspected of working to achieve nuclear weapons capability. By 2003, India, Pakistan, Israel, and North Korea were together believed to stock an estimated 400 nuclear weapons.

- Even as Cousteau and I worked on this book, France reopened its accident-prone "Superphénix" breeder reactor to produce plutonium. But in 1998, the country closed the breeder reactor down for good. By the end of 2006, only 38 percent of the breeder had been disassembled, and this risky and expensive process is not expected to be completed until 2025.

- As we reported, Japan activated its own first fast-breeder reactor, dubbed "Monju" after the god of wisdom, in 1994. Monju closed in 1995 after a serious sodium leak and fire and is not expected to function again until 2008.

- Mutually Assured Destruction depends on the theory that no nation would launch a first strike out of fear of retaliation. As bomb-building abilities proliferate, there is no certainty today that anyone would know who launched a first strike—or that retaliation would be possible, given that the targeting systems originally aimed at the Soviet Union may not function against so-called over-the-horizon countries in the Middle East.

- Even more chilling than an exchange between nuclear-armed nations is the prospect of a terrorist atomic attack, which has become all the more conceivable given the proliferation caused by the lure of lucrative profits on the black market. In our text, Cousteau and I reported that a Pakistani physicist, Abdul Qadeer Khan, had stolen blueprints for weapons production from Holland, where he had been working for a uranium enrichment facility, and that, though condemned in absentia by Holland, he was named head of the nuclear program of Pakistan. He went on to become his country's hero. Pakistan exploded a nuclear bomb in a 1998 test. Yet having betrayed Holland, the country that provided him training, Khan went on to betray his own country as well: In 2004 he confessed that for a decade he had been building a $400-million empire for himself by selling off nuclear goods and nuclear secrets to outlaw regimes. He provided instructions for bomb building to Libya and nuclear hardware to North Korea. It is feared that he provided bomb-building instructions to Iran. He is believed to have had dealings with al Qaeda, Saudi Arabia, Egypt, and many African countries rich in natural supplies of uranium. The day after Khan confessed, Pakistan pardoned him, raising suspicions about the extent to which members of the Pakistani military profited along with him. Worse, the Swiss and German governments did not arrest his cohorts. Khan is said to be ill and to be living as a recluse who is disdained by his countrymen. All the same, while international agents know Khan's every move, they do not know the actions of his still-extant black-market nuclear network.

○ We reported that in 1994, Iran renewed requests to purchase nuclear reactors, and that established nuclear nations, suspecting the country of working on a bomb project, tried to block the sales. In fact, in 2002, informants reported that Iran was enriching uranium, in violation of the Nuclear Nonproliferation Treaty. In December 2006, after the United Nations Security Council imposed economic sanctions on Iran, the country's parliament retaliated by voting not only to accelerate its nuclear research but also to limit its cooperation with the IAEA.

○ In 1994, North Korea agreed to abandon its plutonium reprocessing project, useful for bomb building, in exchange for U.S. diplomatic recognition and light-water reactors that would produce nuclear energy. In 1998, however, North Korea launched medium-range missiles into the Sea of Japan, and in 2002 solid intelligence revealed a North Korean program for uranium enrichment. Given these circumstances, the United States chose not to fulfill its side of the bargain. In 2002, North Korea proclaimed that it once again was undertaking a plutonium-based nuclear program, and by October 2006, North Korea had produced and exploded its first test atomic bomb.

○ In 2006, the Israeli prime minister claimed to have been misinterpreted when he accidentally included his own country in a litany of nations with an atomic arsenal. Iran immediately demanded that the United Nations place Israeli facilities on its inspection list.

○ In March 2007, as negotiators attempted to curtail weapons development in North Korea and Iran, the United States announced that it was building a new generation of nuclear warheads.

○ In December 2006, the U.S. Congress approved a White House offer to recognize India as a sixth declared nuclear power. If accepted internationally, this recognition will not only bestow India with great prestige but also will enable it to buy more nuclear material from the other major powers. For thirty years, India has refused to sign the Nuclear Nonproliferation Treaty and will not be required to do so now.

○ Increasing evidence points to al Qaeda's intention of using an atomic bomb if it can acquire one.

○ In 2006, 60 percent of Americans feared a nuclear attack more than they had a decade before.

* * *

According to legend, there is a way to tell if a stone is an authentic emerald. You heat it to a certain temperature, and if it shatters—it was real.

We have inherited our emerald—our Earth—only in usufruct. If indeed it has become too late to remedy our actions, if we shatter our emerald, we are committing a crime not only against ourselves but against future generations.

The figures indicate that many of our margins for error have disappeared. Some experts say there is still a narrow possibility that we can change our ways and preserve the earth for our children. Their comments bring to mind one of Cousteau's first speeches on environmental activism, which he delivered in the early seventies. He woke up the next morning to find a full-page newspaper report on his speech. The banner headline: "We Have No Time to Lose." Too many years have passed. We must act now before no time is left.

SELECTED BIBLIOGRAPHY

While hundreds of sources were instrumental in the years of research that went into this book, we are especially indebted to the books and articles in the following list. Also included, for further reading, is a selection of earlier works by Cousteau focusing on some of the incidents mentioned in this book.

Boorstin, Daniel. *The Discoverers*. New York: Random House, 1983.

Bundy, McGeorge. *Danger and Survival: Choices About the Bomb in the First Fifty Years*. New York: Random House, 1988.

Clark, Ronald W. *Einstein: The Life and Times*. New York and Cleveland: The World Publishing Company, 1971.

Cousteau, Jacques-Yves. *National Geographic*:

    "Fish Men Explore a New World Undersea," October 1952, vol. 102, no. 4: pp. 431–472.

    "Fish Men Discover a 2,200-Year-Old Greek Ship," January 1954, vol. 105, no. 1: pp. 1–35.

    "To the Depths of the Sea by Bathyscaphe," July 1954, vol. 106, no. 1: pp. 67–79.

    "Diving Through an Undersea Avalanche," April 1955, vol. 107, no. 4: pp. 538–542.

    "*Calypso* Explores for Underwater Oil," August 1955, vol. 108, no. 2: pp. 155–184.

    "Exploring Davy Jones's Locker with *Calypso*," February 1956, vol. 109, no. 2: pp. 149–161.

"*Calypso* Explores an Undersea Canyon," March 1958, vol. 113, no. 3: pp. 373–396.

Cousteau, Jacques-Yves, and Philippe Cousteau. *The Shark: Splendid Savage of the Sea*. New York: Doubleday, 1970.

Cousteau, Jacques-Yves; Philippe Cousteau; and Philippe Diolé. *Three Adventures: Galápagos, Titicaca, The Blue Holes*. New York: Doubleday, 1973.

Cousteau, Jacques-Yves, with Philippe Diolé. *Life and Death in a Coral Sea*. New York: Doubleday, 1971.

Cousteau, Jacques-Yves, with James Dugan. *The Living Sea*. New York: Harper and Row, 1962.

Cousteau, Jacques-Yves, with Frédéric Dumas. *The Silent World*. New York: Harper Bros., 1953.

Dubos, Rene. "The Mysteries of Life." *Propaedia, Encyclopedia Britannica*. 15th ed. Chicago: Encyclopedia Britannica Inc., 1974.

Dyson, Freeman. *Disturbing the Universe*. New York and London: Harper and Row, 1979.

Edgerton, Harold E. "Photographing the Sea's Dark Underworld." *National Geographic*, April 1955, vol. 107, no. 4: pp. 523–537.

Eiseley, Loren. *The Unexpected Universe*. Orlando: Harcourt, Brace and World, 1969.

Eiseley, Loren. "The Cosmic Orphan." *Propaedia, Encyclopedia Britannica*. 15th ed. Chicago: Encyclopedia Britannica Inc., 1974.

Gould, Stephen Jay. *The Panda's Thumb: More Reflections in Natural History*. New York: W. W. Norton and Co., 1980.

Houot, Lieutenant Commander Georges S. "Two and a Half Miles Down." *National Geographic*, July 1954, vol. 106, no. 1: pp. 80–86.

Kaysen, Carl, Robert S. McNamara, and George W. Rathjens. "Nuclear Weapons After the Cold War." *Foreign Affairs*, Fall 1991: pp. 308–309.

Marden, Luis. "Master of the Deep: Jacques-Yves Cousteau 1910–1997." *National Geographic*, February 1998, vol. 193, no. 2: pp. 70–79.

Mumford, Lewis. "Gentlemen: You Are Mad!" *Saturday Review of Literature*, March 2, 1946, vol. XXIX: pp. 5–6.

Perkovich, George. "The Plutonium Genie." *Foreign Affairs*, Summer 1993: pp. 153–165.

Piel, Gerard. *Science in the Cause of Man*. New York: Knopf, 1962.

Ritchie-Calder, Lord. "Knowing How and Knowing Why." *Propaedia, Encyclopedia Britannica*. 15th ed. Chicago: Encyclopedia Britannica Inc., 1974.

Shulman, Seth. "When a Nuclear Reactor Dies, $98 Million Is a Cheap Funeral." *Smithsonian*, October/November 1989: pp. 54–69.

Smith, Gerard, and Helena Cobban. "A Blind Eye to Nuclear Proliferation." *Foreign Affairs*, Summer 1989: pp. 53–70.

Von Hippel, Frank, Marvin Miller, Harold Feiveson, Anatoli Diakov, and Frans Berkhout. "Eliminating Nuclear Warheads." *Scientific American*, August 1993: pp. 44–49.

Weart, Spencer R. *Scientists in Power*. Cambridge, MA: Harvard University Press, 1979.

Wiesner, Jerome. *Where Science and Politics Meet*. New York: McGraw Hill, 1961.

Born in 1910, JACQUES COUSTEAU was world renowned as an ocean explorer, filmmaker, educator, and environmental activist. He won three Oscars and the Palme d'Or for his films, was nominated for forty Emmys during the run of his TV series *The Undersea World of Jacques Cousteau* and *The Cousteau Odyssey*, and wrote or coauthored more than seventy-five books, including *The Silent World*, which has sold more than five million copies in twenty-two languages. As director of the Oceanographic Museum of Monaco and a director of the Marine Radioactivity Laboratory of the International Atomic Energy Agency, he was active in the conservation and anti-nuclear-proliferation movements. He died in 1997.

SUSAN SCHIEFELBEIN has won the National Magazine Award and the Front Page Award for her cover stories on social issues. A former senior editor at the *Saturday Review*, where she first worked with Cousteau, she went on to write the narration for many of his documentary films, including winners of the Peabody and the Ace. She lives in Paris.